Mastering Wireshark 2

Develop skills for network analysis and address a wide range
of information security threats

Andrew Crouthamel

BIRMINGHAM - MUMBAI

Mastering Wireshark 2

Commissioning Editor: Vijin Boricha
Acquisition Editor: Prachi Bisht
Content Development Editor: Trusha Shriyan
Technical Editor: Sayali Thanekar
Copy Editor: Laxmi Subramanian
Project Coordinator: Kinjal Bari
Proofreader: Safis Editing
Indexer: Tejal Daruwale Soni
Graphics: Jisha Chirayil
Production Coordinator: Shraddha Falebhai

First published: May 2018

Production reference: 1290518

Published by Packt Publishing Ltd.
Livery Place
35 Livery Street
Birmingham
B3 2PB, UK.

ISBN 978-1-78862-652-1

www.packtpub.com

`mapt.io`

Mapt is an online digital library that gives you full access to over 5,000 books and videos, as well as industry leading tools to help you plan your personal development and advance your career. For more information, please visit our website.

Why subscribe?

- Spend less time learning and more time coding with practical eBooks and Videos from over 4,000 industry professionals

- Improve your learning with Skill Plans built especially for you

- Get a free eBook or video every month

- Mapt is fully searchable

- Copy and paste, print, and bookmark content

PacktPub.com

Did you know that Packt offers eBook versions of every book published, with PDF and ePub files available? You can upgrade to the eBook version at `www.PacktPub.com` and as a print book customer, you are entitled to a discount on the eBook copy. Get in touch with us at `service@packtpub.com` for more details.

At `www.PacktPub.com`, you can also read a collection of free technical articles, sign up for a range of free newsletters, and receive exclusive discounts and offers on Packt books and eBooks.

Contributor

About the author

Andrew Crouthamel is an experienced senior network engineer and IT trainer who resides in Doylestown, PA, and currently works with organizations including NASA, ESA, JAXA, Boeing, and the US Air Force. His passion for teaching is reflected in his work, which is filled with excitement and real-world anecdotes.

Packt is searching for authors like you

If you're interested in becoming an author for Packt, please visit `authors.packtpub.com` and apply today. We have worked with thousands of developers and tech professionals, just like you, to help them share their insight with the global tech community. You can make a general application, apply for a specific hot topic that we are recruiting an author for, or submit your own idea.

Table of Contents

Preface

Wireshark, a combination of Kali and Metasploit, deals with the second to the seventh layers of network protocols. The book will introduce you to various protocol analysis methods and teach you how to analyze them. You will discover and work with some advanced features, which will enhance the capabilities of your application. By the end of this book, you will be able to secure your network using Wireshark 2.

Who this book is for

If you are a security professional or a network enthusiast who is interested in understanding the internal working of networks and have some prior knowledge of using Wireshark, then this book is for you.

What this book covers

Chapter 1, *Installing Wireshark 2*, teaches you how to install Wireshark on Windows, macOS, and Linux.

Chapter 2, *Getting Started with Wireshark*, tells you about what's new in Wireshark 2. It will also teach you how to capture traffic and how to save, export, annotate, and print packages.

Chapter 3, *Filtering Traffic*, teaches you about BPF syntax and how to create one. It further explains how to use BPF to apply it as a capture filter and reduce the packets, how to create and use display filters, and how to follow streams—both TCP and UDP.

Chapter 4, *Customizing Wireshark*, explains how to apply preferences in Wireshark and customize them. You will learn how to create profiles for different analysis requirements.

Chapter 5, *Statistics*, provides an overview of TCP/IP and time values and summaries. You will also take a look at the expert system usage feature of Wireshark.

Chapter 6, *Introductory Analysis*, explains the basics of DNS and some DNS query examples. You will also learn about ARP resolution and how to resolve an IP address to a physical MAC address on an Ethernet bus. You will also acquire knowledge about IPv4 and IPv6 headers, the flags within them, and the fragmentation.

Chapter 7, *Network Protocol Analysis*, teaches you about UDP analysis: the connectionless protocol, TCP analysis: the connection-oriented protocol, and finally, graphing I/O rates and TCP trends: visualization of the data analyzed.

Chapter 8, *Application Protocol Analysis I*, talks about HTTP, both in an unencrypted fashion and an encrypted fashion, and how to decrypt that. You will also look into FTP in all of its many flavors, including active mode, passive mode, and the encrypted flavors of FTPS and SFTP.

Chapter 9, *Application Protocol Analysis II*, teaches you email analysis using POP and SMTP. We will also look at VoIP analysis using SIP and RTP.

Chapter 10, *Command-Line Tools*, teaches you how to run Wireshark from the command line, tshark, tcpdump, and running dumpcap.

Chapter 11, *A Troubleshooting Scenario*, covers troubleshooting a specific issue within Wireshark.

To get the most out of this book

You will need to have Wireshark installed in a Windows/Linux/macOS system.

Download the color images

We also provide a PDF file that has color images of the screenshots/diagrams used in this book. You can download it here: https://www.packtpub.com/sites/default/files/downloads/MasteringWireshark2_ColorImages.pdf.

Conventions used

There are a number of text conventions used throughout this book.

CodeInText: Indicates code words in text, database table names, folder names, filenames, file extensions, pathnames, dummy URLs, user input, and Twitter handles. Here is an example: "So, pcapng is the next generation of the pcap file extension."

Any command-line input or output is written as follows:

```
nslookup wireshark.org 8.8.8.8
```

Bold: Indicates a new term, an important word, or words that you see onscreen. For example, words in menus or dialog boxes appear in the text like this. Here is an example: "We can see the target in the **Address Resolution Protocol (request)** option."

 Warnings or important notes appear like this.

 Tips and tricks appear like this.

Get in touch

Feedback from our readers is always welcome.

General feedback: Email `feedback@packtpub.com` and mention the book title in the subject of your message. If you have questions about any aspect of this book, please email us at `questions@packtpub.com`.

Errata: Although we have taken every care to ensure the accuracy of our content, mistakes do happen. If you have found a mistake in this book, we would be grateful if you would report this to us. Please visit `www.packtpub.com/submit-errata`, selecting your book, clicking on the Errata Submission Form link, and entering the details.

Piracy: If you come across any illegal copies of our works in any form on the Internet, we would be grateful if you would provide us with the location address or website name. Please contact us at `copyright@packtpub.com` with a link to the material.

If you are interested in becoming an author: If there is a topic that you have expertise in and you are interested in either writing or contributing to a book, please visit `authors.packtpub.com`.

Reviews

Please leave a review. Once you have read and used this book, why not leave a review on the site that you purchased it from? Potential readers can then see and use your unbiased opinion to make purchase decisions, we at Packt can understand what you think about our products, and our authors can see your feedback on their book. Thank you!

For more information about Packt, please visit `packtpub.com`.

Installing Wireshark 1

In this chapter, we'll cover the following topics:

- Installation and setup
  - Installing Wireshark on Windows
  - Installing Wireshark on macOS
  - Installing Wireshark on Linux

Installation and setup

In this section, we'll take a look at installing Wireshark on Windows and installing Wireshark on macOS and Linux.

Installing Wireshark on Windows

You will need to perform the following steps:

1. Go to the `https://www.wireshark.org/` web page:

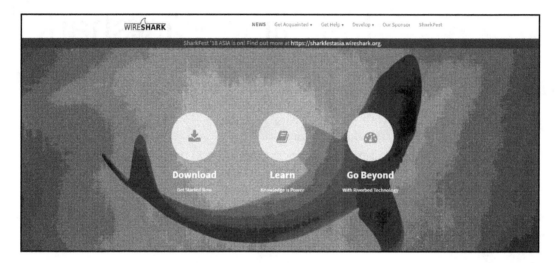

2. When you get there, scroll down on the home page and click on **Download**.
3. The latest version of Wireshark will be visible. Select the installer for the version of Windows that you are currently running.

   Most people on a modern computer, on a modern version of Windows, will be running 64-bit.

   If you happen to be running an older version of Windows on 32-bit or older hardware, make sure you select the 32-bit version. If you're not sure which one to use, do the following:

   Open your **Control Panel**, go to **System and Security**, and click on the **System** link. In the **System** section, you'll see that it says **System type: 64-bit Operating System**. If you have a 32-bit, it'll show that here as well.

4. Back on the download Wireshark page, download the version that you need, and run that file; now, click on **Next** to begin the setup.

5. Read the **License Agreement**, click on **I Agree**, and select the features of Wireshark that you wish to include. Most people include all of the defaults. You'll see here that we have the main Wireshark application; we have the classic interface version of Wireshark; and we have **TShark**, which is a command-line version of running Wireshark; as well as some plugins, the **User's Guide**, and some additional tools:

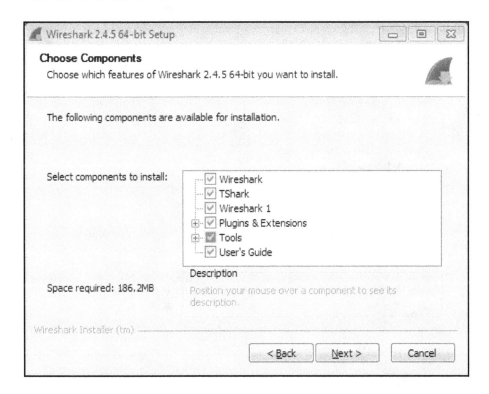

6. We'll go ahead and accept the defaults, and then click on **Next**. And, in this window, we can go ahead and customize what shortcuts show up and whether file extensions are associated to Wireshark. We're going to turn off the **Wireshark Legacy Quick Launch Icon** and **Wireshark Legacy Start Menu Item**:

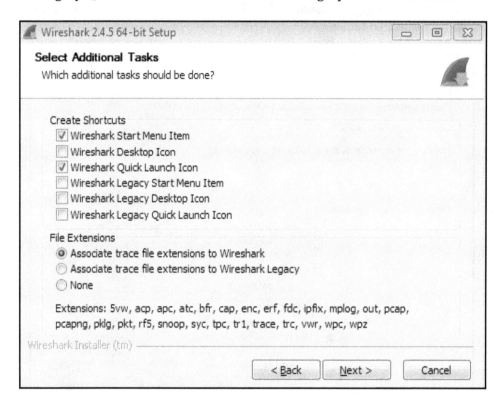

7. Go ahead and click on **Next**, and select a location for Wireshark to install. We'll select the defaults here as well.

8. And on the next page here, it says: **Install WinPcap?** If you don't have WinPcap installed, leave that checked and it will install it as part of the install process:

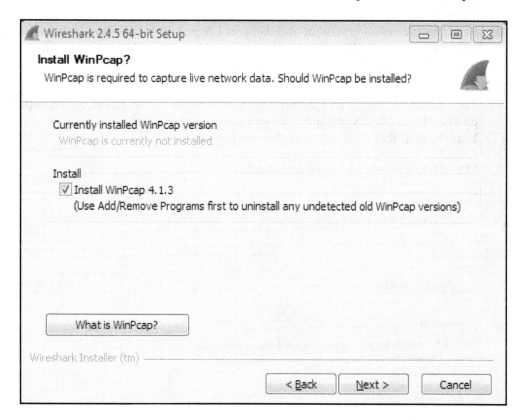

 WinPcap is the driver that allows Wireshark to interface with your network card. WinPcap is required for it to be able to view all the packets.

9. Click on **Next**.

 You can also install USBPcap, which allows you to view the traffic on a USB connection. Most people won't need this, so we'll leave that unchecked.

10. Go ahead and click on **Install** and Wireshark will now install.
11. Partway through the install, the WinPcap installer will then run, and we'll go ahead and click on **Next**.
12. Read the **License Agreement** and click on **I Agree**. You can then decide whether or not you want the WinPcap driver to run at boot time. Most people allow it to do so. We'll leave that as default, and click on **Install**.
13. That will finish very quickly; then click on **Finish**.
14. The Wireshark install will then continue. When the text window says completed, go ahead and click **Next**; and then you can select whether or not you want to run Wireshark at that moment, and click on **Finish**.

Once the Wireshark GUI loads up, you are done:

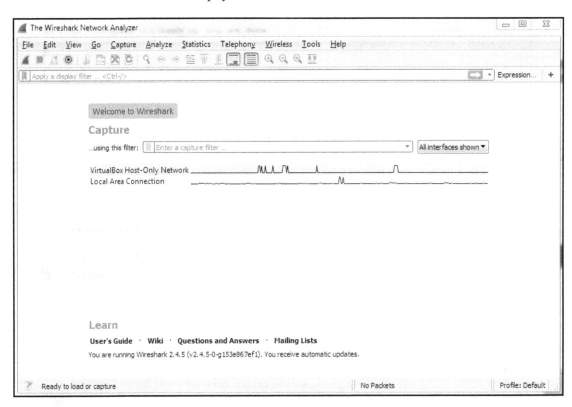

In the next section, we'll go over how to install Wireshark on macOS and Linux.

Installing Wireshark on macOS

To install Wireshark on macOS, perform the following steps:

1. Start by going to the `https://www.wireshark.org/` web page.
2. When you're on the web page, scroll down on the main page, and click on **Download**. The latest version of Wireshark will be displayed.
3. Go down to **macOS 10.6 and later Intel 64-bit .dmg**, and click on it to download:

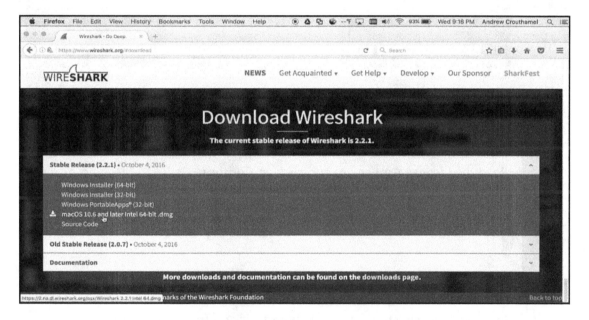

4. At this point, we can choose to save the file to our **Downloads** folder and then open it, or simply open it directly off the web page with the **DiskImageMounter (default)**.
5. Go ahead and click on **OK**. It downloads the file and opens it up.

6. We can then double-click on the `PKG` file, and click on **Continue**:

7. Read the **License Agreement** and click on **Continue** again, and then on **Agree** to indicate that you agree to the license agreement.

If you wish to change the install location, you can do so now.

8. Click on **Install**.
9. Enter your administrator credentials and click on **Install Software**.
10. Once the installation is successful, click on **Close**.

11. If you go to your applications list in the lower right, and scroll down, you should see Wireshark at the bottom of the list. You can select Wireshark, and you can see that it's now loaded:

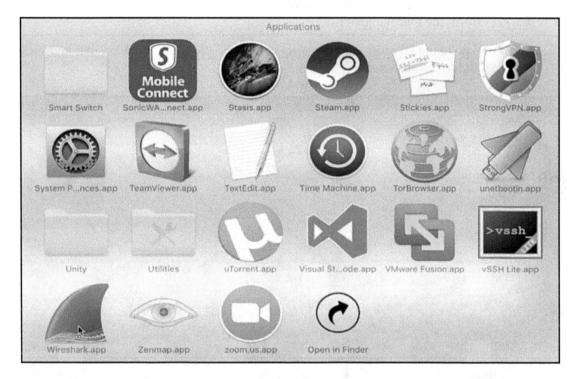

Once Wireshark is up and running, that's it—you're done.

Installing Wireshark on Linux

Installing Wireshark on Linux will differ, based on the distribution that you're using. Here, I'm using one of the most common distributions available: Ubuntu. In order to install Wireshark, perform the following steps:

1. We'll go to the Ubuntu software application; go ahead and click on that and we'll search for synaptic:

 Synaptic is a package manager similar to the Ubuntu software application, but it gives you more control.

2. Simply click on **Install**; enter your administrator password (your root password) and the software will be installed:

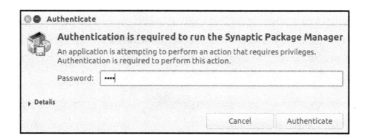

3. Go ahead and click on **Synaptic Package Manager** to open that up. Enter in our credentials again, and now we have our Synaptic application loaded:

4. This is again very similar to the Ubuntu software application, but it's less pretty.
5. Click on the **Search** button, and we will search for Wireshark. Enter `wireshark` and click on **Search**, and you'll see everything that has Wireshark in its name or description now shows up in the package list:

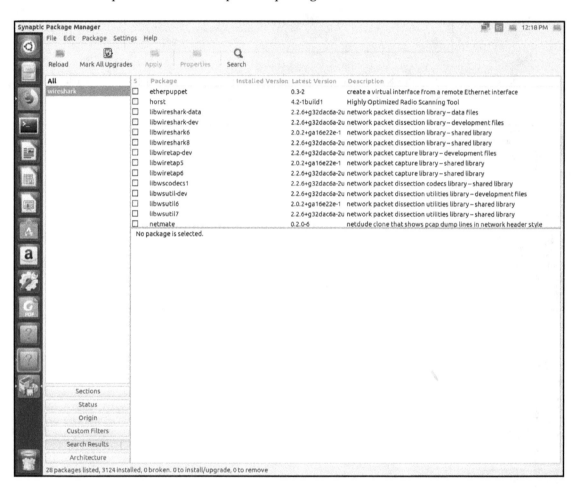

6. We'll scroll down and select the main **wireshark** package, just the one that says Wireshark, as shown in the following screenshot:

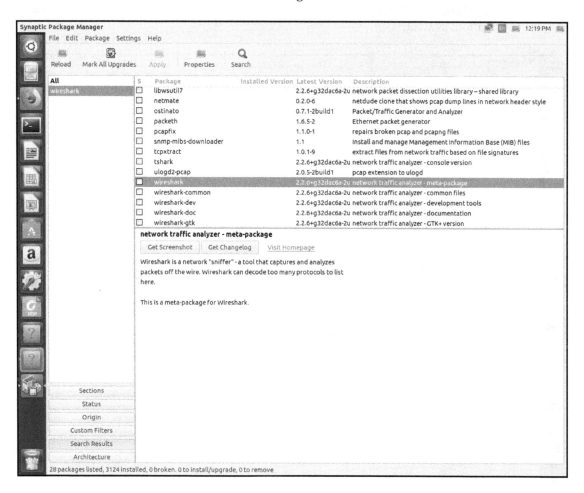

7. So we'll select that; click on that, and select **Mark for Installation**. It will then ask you if it's okay to install other packages that are required. We can say sure, that's fine; **Mark** them for installation as well. So now all of our dependencies will be installed, as well:

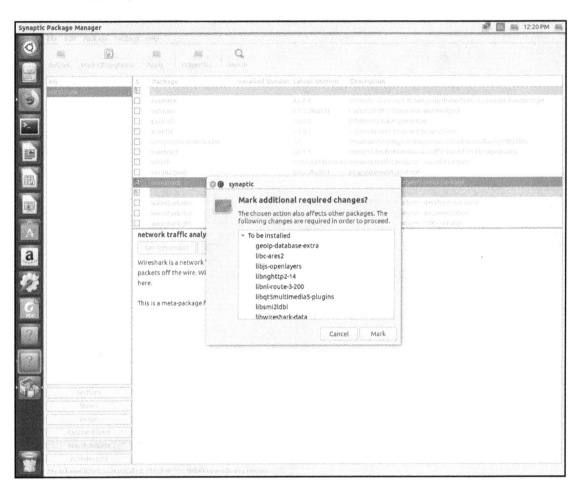

8. We can then go up and click on **Apply**, and it'll tell us that we'll be installing the following packages. Click on **Apply** again, and Synaptic will go ahead and download and install all the programs that we've selected:

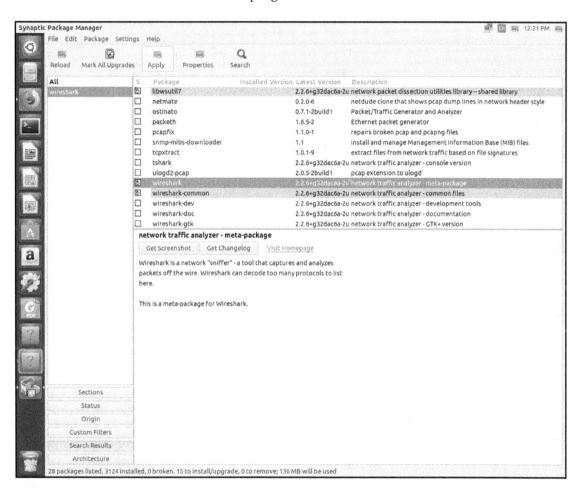

At this point, you'll receive a window popup asking you if non-superusers should be able to capture packets. I would recommend selecting **Yes**. That will allow you to separate your root user from your standard users, but still allow you, as a standard user, to capture packets.

9. Once everything is complete, you'll receive a **Changes applied** window. It'll say: **Successfully applied all changes. You can now close the window.**

10. Simply click on **Close**, and you'll see everything here marked in green is now installed, including Wireshark:

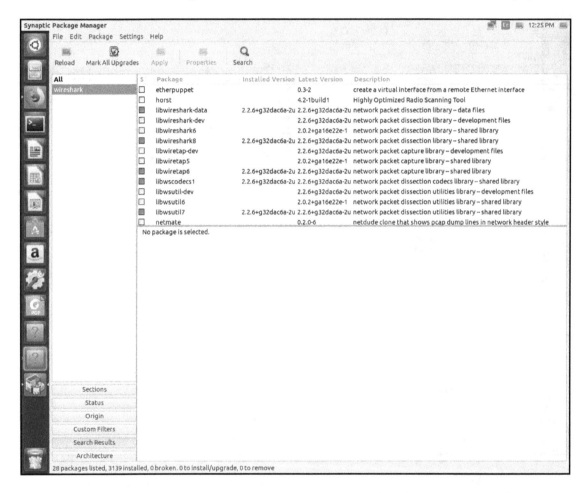

11. So, at this point, we can close this program as well as the Ubuntu software. Next, click on the **Search** button in the upper left corner of the interface, and we'll type in wireshark. It automatically shows that Wireshark is here. We can simply click on that and it will load Wireshark.

12. At this point, once Wireshark loads, you're done.

Summary

In this chapter, you've learned how to install Wireshark on both macOS and Linux—specifically, Ubuntu.

In Chapter 2, *Getting Started with Wireshark*, we are going to take a look at what's new in Wireshark 2, capturing traffic, saving and exporting packets, annotating and printing packets, remote capture setup, and remote capture usage.

Getting Started with Wireshark 2

In this chapter, we'll cover the following topics:

- What's new in Wireshark 2
- Capturing traffic
- Saving and exporting packets
- Annotating and printing packets
- Remote capture setup
- Remote capture usage

What's new in Wireshark 2?

There's a new version of Wireshark out—a new major version that has many interesting features. Here, you can see the new Qt GUI:

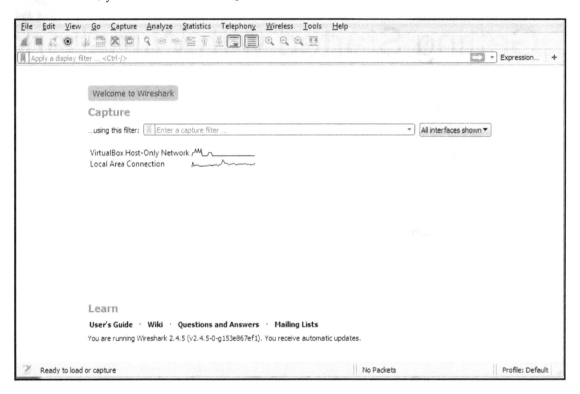

It looks very similar to the Legacy GTK GUI, with few minor tweaks. The main menu bar here has had some icons changed and removed; the general interface is a little bit cleaner. All the general functionality, though, is all the same. Capture options are on the upper left-hand side and they are denoted by a gear icon. When you click on the gear icon, you have multiple tabs for **Input** options, **Output** options, and general **Options**:

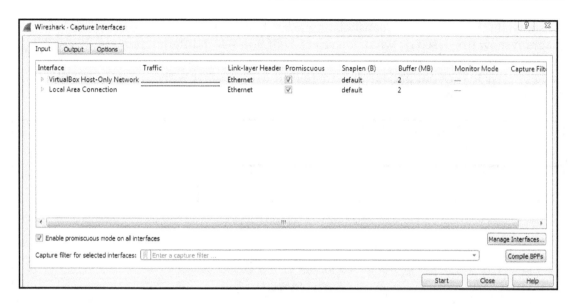

When you click on **Edit | Preferences...**, you can see the preferences window, as shown in the following screenshot. Options such as **Show up to** makes it easy to navigate and view what you need to see:

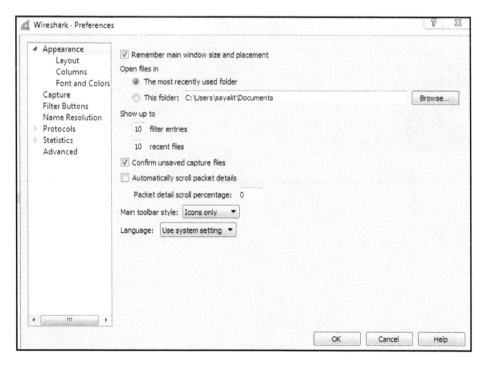

As shown in the following screenshot, on the left-hand side, you can see the related packets diagram show up, based on what you select. So if you select different packets, this will change in size and shape; and what might appear for you is then what you select. This makes it easy to pick out packets that are related to each other without having to follow TCP or UDP streams:

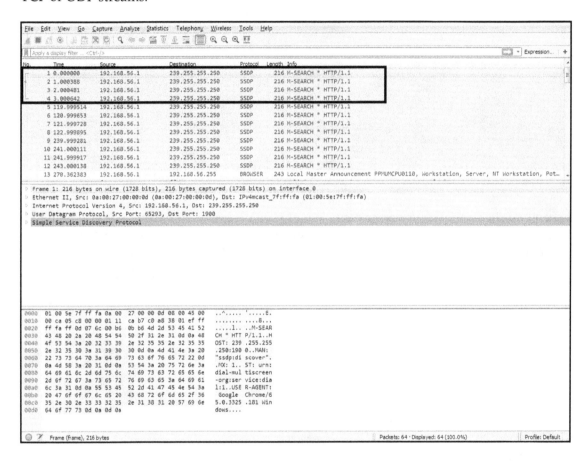

Under the **Statistics** menu that is present in the menu bar, many of these statistics options now have a similar-looking window, as shown in the following screenshot. If you look at how the buttons, filters, and general interface is set up, they're all now standardized and look very, very similar to each other, which I'm sure makes coding much easier for those who work on the Wireshark code:

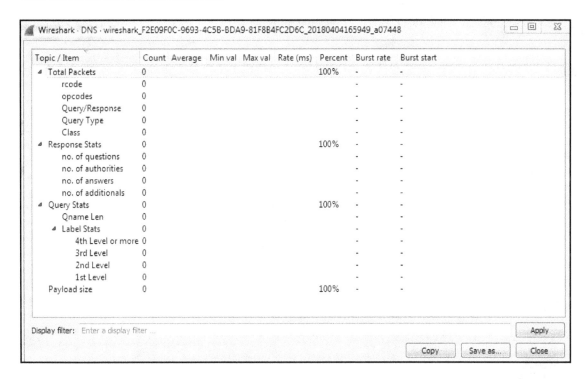

Topic / Item	Count	Average	Min val	Max val	Rate (ms)	Percent	Burst rate	Burst start
◢ Total Packets	0					100%	-	-
rcode	0						-	-
opcodes	0						-	-
Query/Response	0						-	-
Query Type	0						-	-
Class	0						-	-
◢ Response Stats	0					100%	-	-
no. of questions	0						-	-
no. of authorities	0						-	-
no. of answers	0						-	-
no. of additionals	0						-	-
◢ Query Stats	0					100%	-	-
Qname Len	0						-	-
◢ Label Stats	0						-	-
4th Level or more	0						-	-
3rd Level	0						-	-
2nd Level	0						-	-
1st Level	0						-	-
Payload size	0					100%	-	-

Display filter: Enter a display filter ...

Apply Copy Save as... Close

Click on **Statistics** | **I/O Graph**; now you can see the Wireshark IO graph. In the bottom left-hand, you can click on the plus icon and add multiple items to the chart on your IO graph, and you can do this an unlimited number of times:

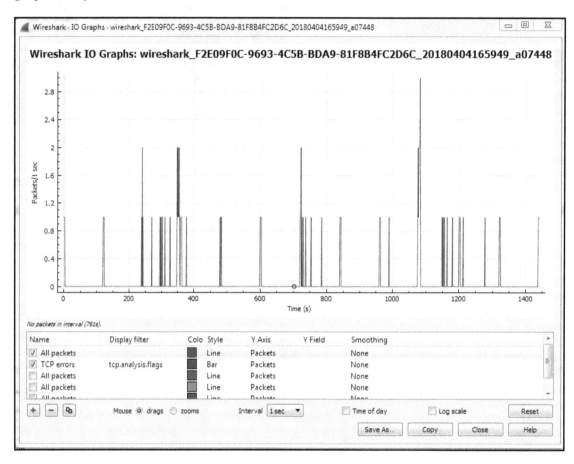

Additionally, any changes you make in here are saved to your profile. With this graph, you can also click on **Save As...** and select different file formats to choose from:

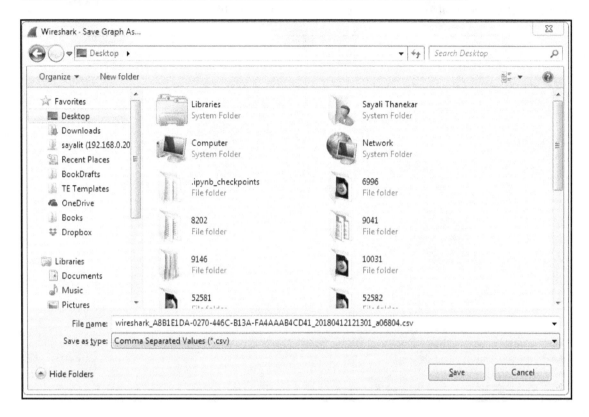

Click on **Analyze | Follow | UDP Stream**; you can see the follow stream dialog box has been updated so that it now allows you to select whether it's the entire conversation or just one side at a time. It also allows you to search for text within:

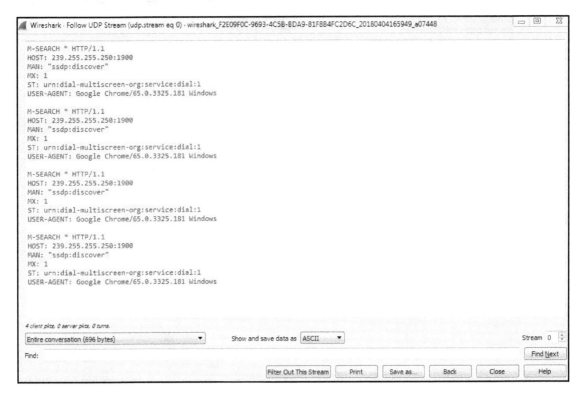

In the preceding screenshot you can see the context-aware hints in action. Within this stream, if you look at the bottom, you have some information such as client packets, server packets, and so on, that changes based on what you're hovering over. The main capture window will change to that actual packet.

This is very handy for jumping through the data and being able to see it in relation to the entire capture.

Let's now see how we'll capture traffic and get the first packets in that main window.

Capturing traffic

One of the first things I'm sure you want to do in Wireshark is to begin capturing some traffic so that you can get used to the utility and possibly diagnose some issues on your own network. In this section, we'll talk about exactly that: where to capture that traffic and how to capture it.

Wireshark needs to receive packets in one way or another, so that you may begin analyzing the data and performing your network diagnostics. There are several ways of doing so in Wireshark. One way is to begin capturing on a local device with Wireshark installed through the GUI. You also have the option of doing so through a command-line. You can capture remotely from a Wireshark install on a management computer, for example. It can retrieve the packets being received and sent from a device somewhere else on your network, using a special driver install. You can also capture the traffic inline on the wire, which means you place a device called a **test action port** (**TAP**) somewhere along the data path that you need to diagnose, and it will then send that data back to your diagnostic utilities, one of which could possibly be Wireshark. And lastly, we'll go over how to store packets locally on a internetwork device (specifically, a Cisco router or switch) for export into Wireshark as a `pcap` file.

How to capture traffic

In order to capture traffic inline for Wireshark, you need to place some sort of device on the wire where it can see the traffic being sent and received, and then replicate that traffic to additional ports for your diagnostic machines, which might be possibly running Wireshark, for example. One of the early devices that we can use for older networks that we're running half duplex is the hub. This is the predecessor to the switch, and it has a very basic functionality where it sees the electrical signals being sent across the wire, and it replicates those electrical signals out all the other ports that it has, without any care as to what's on these actual ports. It's just a splitter, basically. That's great for a slower, older, half-duplex network; but for a modern, switched, full-duplex network, you'll need something a little bit fancier. One of the devices that you could use is a TAP.

There are four different TAPs available:

- Non-aggregating TAPs
- Aggregating TAPs
- Regenerating TAPs
- Link aggregation TAPs

Each one of these TAPs have different functions. I mentioned **switched port analysis** (**SPAN**) ports or port mirroring. In a modern-switched network, this is a very common way of receiving traffic. If you have a managed switch, such as a Cisco switch or whoever's it might be, you can go into the switch and tell it to replicate the traffic that it sees on one port to a different port. This port could then be connected to your Wireshark machine to capture traffic. It's very useful for modern networks because there's no other hardware required. You can just go into the switch and tell it to replicate the data out to your monitoring system. In order to capture traffic on wireless, you need to be aware that there are multiple modes that you could use. There are two modes that we will be discussing:

- **Monitor mode**: This mode receives all packets on a specified channel. So, in the US we have 11 channels on 2.4 GHz, for example. You could tell your network card or wireless card to receive all traffic on channel number 3, and then it would capture all of that traffic for any SSID and any network that is on channel 3.
- **Promiscuous mode**: This mode is more common to find in your wireless drivers, and it allows you to receive all packets on a connected SSID, on a connected network. If you're connected to your work network or your home network-whatever it is you're trying to diagnose-it'll capture anything that's traversing that network name and that SSID. But it will ignore any others on the same channel, and it will certainly ignore anything else on any other channel, as well.

In the following screenshot, we can see that Wireshark is running. You can see that I have a list of interfaces here, including a local area connection and some virtual adapters. I do not have any wireless adapters on this computer, or else they would show up here as well. And any other additional **network interface controller** (**NIC**) cards that you might have-wired cards, it doesn't matter-they'd all show up here in a list:

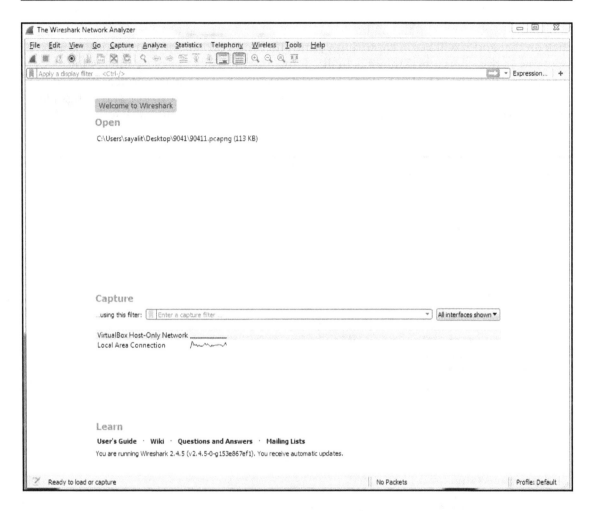

You will also see that there's a chart that's continuing to be drawn by Wireshark, and it's showing us the amount of data that it sees on each connection. This is actually pretty useful, especially if you have a diagnostic computer that has many different interfaces—the different SPAN ports, or whatever it might be. Maybe you turned on SPAN to a specific port that's receiving a lot of data, and you don't know which one it's connected to on the monitoring system. You could take a look here. Whichever port is receiving the most data or the expected amount of data might be the one that you want to try and capture on. So I find that useful on, for example, crowded systems.

In order to capture traffic, all you have to do in the latest version of Wireshark is double-click on that and it will begin capturing your traffic, and you can see that traffic begins to scroll by. In this computer, I'm not actually doing anything which is very interesting, considering how much traffic is being sent and received, but there are services that are running in the background and there's possibly minimized web browsers, and things like that. But you'll see there's quite a bit of communications just on a standard, idling computer:

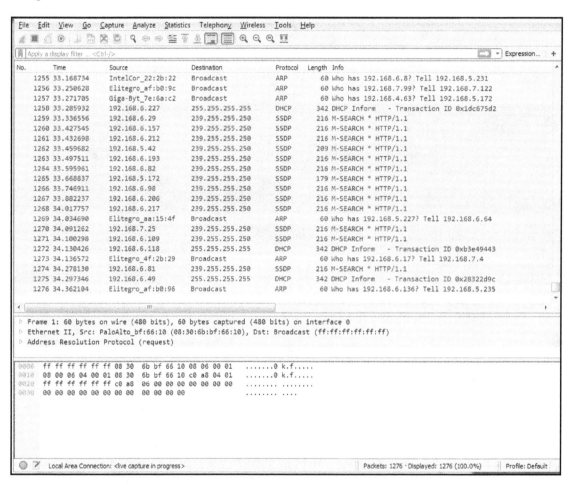

In order to stop this capture, you just go up to the top and click on the stop icon:

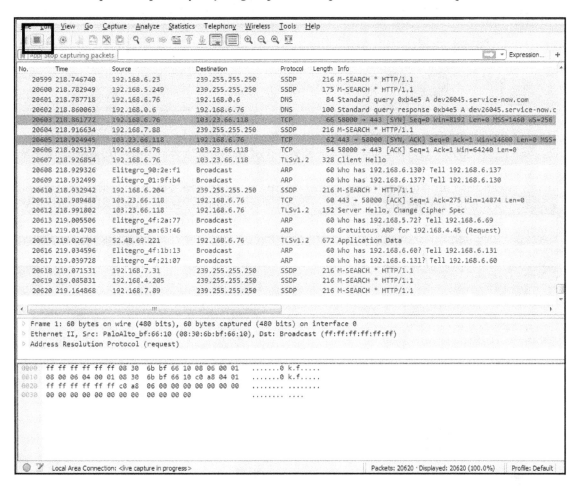

You'll notice that the packets were scrolling by and being updated in real time. Well, this is useful for some situations—it might not be useful for all. So, if you have a system that's receiving a lot of data, for example, possibly gigabits per second or if you're trying to run this on a computer that's very old and slow, that might not be an ideal situation, especially if you're using the GUI.

So you can turn that off so that it doesn't use the graphics card and processor power to try and update this screen for you in real time. In order to do that, perform the following steps:

1. Click on the gear icon, as shown in the following screenshot:

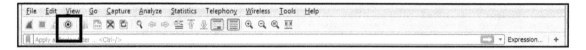

2. Go to **Options**, and you can see that there are some check boxes here that we can turn off. So, you can see the **Update list of packets in real-time**. If I uncheck that, it will prevent the list from populating as it continues to receive packets, and I can turn off **Automatically scroll during live capture**. You will notice that the scroll bar on the right went down to the very bottom. If I turn off **Automatically scroll during live capture**, it would remain up at the top. So these two things are very helpful to disable if you are running on an older computer, like I mentioned:

3. You can also select multiple interfaces. If you go back up to that **Options** selection and you look at the **Input** tab, you can select multiple interfaces with the *Shift* key; or, with the *Ctrl* key, you can select them individually and then click on **Start**, and it will be then capturing on all the interfaces that you selected. Depending on your situation, that may be a useful feature.

In this section, we went over some different ways of capturing packets; how to get them into your Wireshark capturing system. Up next, we will save those packets and export them in various ways.

Saving and exporting packets

In this section, we'll take a look at the following subtopics:

- How to save packet captures
- How to save selected sections of packets, individual packets, and ranges of packets
- How to export packets into other formats
- How to export raw packet data from the capture that you selected

Now that we have Wireshark up and running, let's capture some traffic. We'll select the **Local Area Connection**, and we could either double-click as I mentioned or we'll start the capture up at the top. And we will have some packets coming in. So now, if I want to save this capture (the entire capture—all the packets that I just captured) I'll go to **File | Save As...**; and from here, I can simply choose a filename. So, we'll call it `packets`. And you'll see in the **Save as type**, I chose the file as `pcapng`:

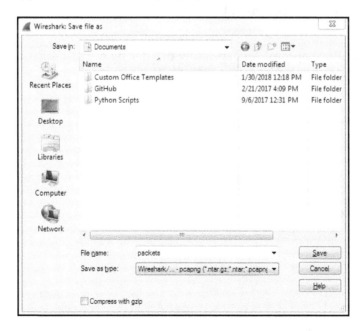

So, pcapng is the next generation of the pcap file extension. It was released with Wireshark 1.8, so it is relatively new and includes some additional features, which we'll get into in future sections. But you should know that the .pcapng file extension is the new standard, so if you see a .pcap file with no ng at the end, that's an older capture file, and you can certainly save it as .pcapng going forward since that's now the default. Just be aware, though, that if you take a pcapng file and save it as an original pcap file, you'll lose some of that functionality that comes with the ng format. So, my recommendation is to stick with the ng format. Almost all plugins and additional software that utilizes pcap files now support the ng format, so you might as well just use that going forward. You'll also notice long, different capture extensions here. You have .gz listed and with the pcapng you also have .gz, ntar.gz, and so on. In order to get that, you have to select the **Compress with gzip** option. So when you select **Compress with gzip** what that will do is, just like putting files into a ZIP file, it will take your capture file and try to compress it to make it smaller. So if it's a large capture, remember a packet capture includes all the data that's traversed your network from your network card that you're capturing on. So, if you're transferring a lot of data at the time that you're doing a capture, all of that will be saved in your capture. It's going to be a 1:1 ratio of the data that's been transferred, so it could be very large. Gzipping that might make sense to you, because then it would be a much smaller file on your hard drive. Additionally, if you're trying to transfer the file across to your network, then that could potentially save time with trying to transfer the file since it would be smaller to transfer. Most of the time it's not used, though, especially if you filter out what you only need to see in a capture and you end up saving only what you require, then they're usually very small.

Now, we will **Save** that capture. And speaking of filtering out just what you want to see and making a capture smaller, let's do exactly that.

So here we have some HTTP traffic, and we'll right-click on that and then click on **Follow | TCP Stream**. That way, we have some sort of stream here that's filtering out all of the other data that's in our packet capture. So we've got seven packets selected. If I want to save just these packets into a new file, I'll go to **File | Export Specified Packets....** We'll call this `packet small` and you'll see here that we have a radio button to select between **Displayed** and **Captured**:

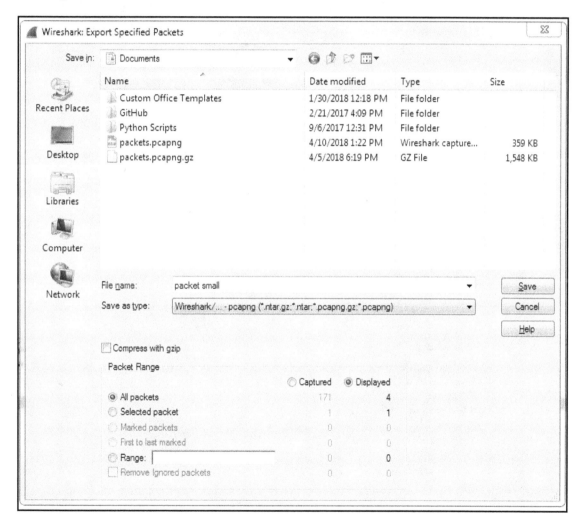

So **Captured** is the entire packet capture. This adds almost 2870 packets. **Displayed** is only going to save what's currently filtered and what's displayed in my packet list view. So if I have **Displayed** selected in all packets, that's going to export all seven of these packets into a new file. Additionally, I could select **Selected packet**. So you'll see here that I have packet number 16 currently selected; it's a slightly different color here. If I choose **Selected packet**, this will only save that one single packet into a new file:

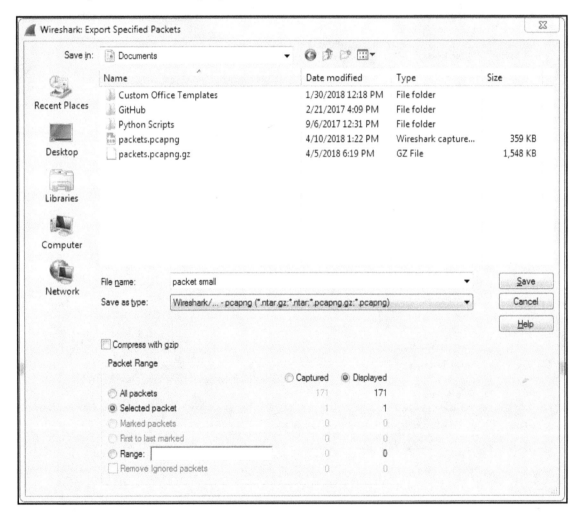

I could also select a **Range**. Now, the range wouldn't show anything right now because we have our own little filter going, but what I can do is clear out this filter. And we'll go back to **Export Specified Packets...** and save the packets range. We could say packets 5 through 200. So there are 196 packets that will be saved into the packets range file:

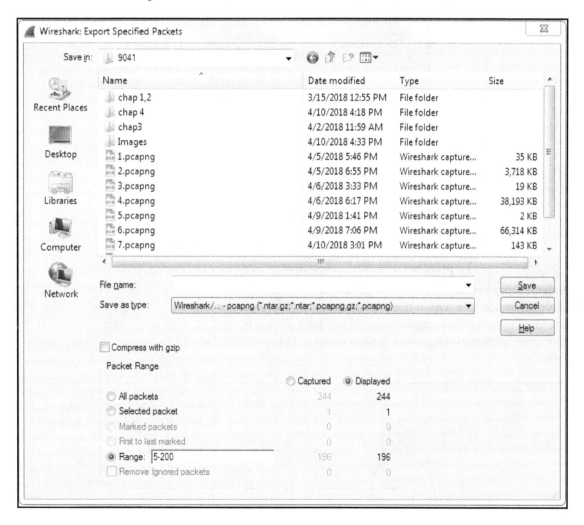

Additionally, you can export your packet dissections by going to **File | Export Packet Dissections** and then choose a format you'd like:

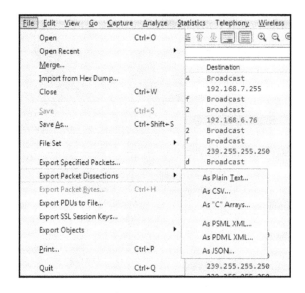

So we'll choose **As Plain Text...**, and we'll call it `packet dissect`. We'll do just the **Selected packet**, and so here we have our `packet dissect` text file, which you can see has the packet number, when the packet came in, source and destination IP addresses, what the protocol was, any information about the protocols within it, and then the basic information that you can see in the packet details section; this is all now saved in the text file as shown here:

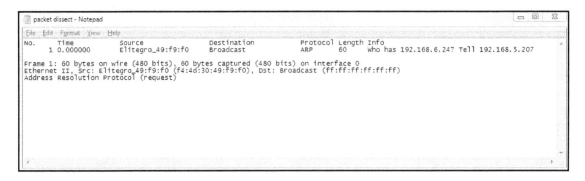

If your packet capture happens to have captured any secured traffic such as any HTTPS, SSL, or SSH traffic, you can use **File** | **Export SSL Session Keys...**, and then save these SSL keys for future use in some other application, if you wish.

You might have noticed one additional export that's grayed out: **Export Packet Bytes...**:

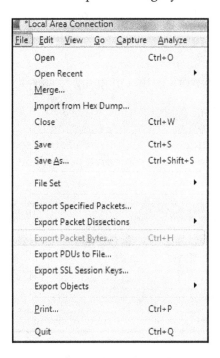

In order to get that to become selectable, you need to select the application data within your packet. So here, we've selected our HTTP packet data. If we go back to **File**, you'll see that **Export Packet Bytes...** is now selectable. If I select that, I can now export my data from my packet in a raw binary format. It is potentially useful if you're doing something with a hex editor or combining these pieces together for some other application.

In this section, you learned how to save and export packets: the entire packet capture, subsets of that packet capture such as filtered views, individual packets, as well as exporting raw data into different formats such as text files.

Annotating and printing packets

In this section, we'll take a look at the following subtopics:

- How to use the new annotation feature, also known as comments
- How to find packets that have annotations, and there are multiple ways of doing so
- How to print packets

Now, let's get some packets to work with. I'm going to start a quick capture.

To create comments for a packet capture, the entire capture itself, you can do so in the bottom left-hand corner of Wireshark. You see there are two icons down there: one's a circle icon called the expert information we'll get into in a future section, and there's a pencil with a packet capture icon. If you do a mouseover on a pencil icon, it'll say **Open the Capture File Properties dialog**:

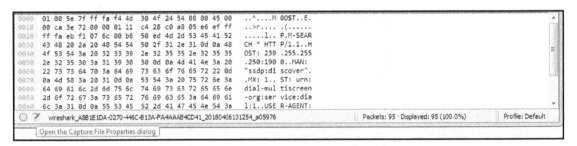

Click on that and it will open up a pop-up window that gives you a bunch of different information for the capture itself. And there's a bottom section here that says **File Comment**, and here you can enter whatever sort of description you want for the entire packet capture. So maybe this is, `Capture from the management PC to the server. Data appears slow.`. Click on the **Save Comments** button, and this will save the comments for you:

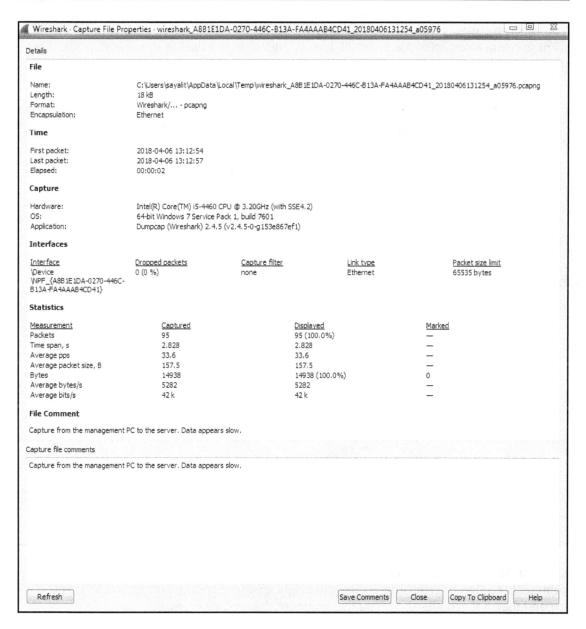

You'll also see that the **File Comment** appeared at the bottom of that top pane when I clicked on **Save Comments**, as well.

You can also create comments for individual packets, and to do so you select the packet you want to create a comment for, right-click on it, and go to the **Packet Comment....** You'll see there's a little pop-up window for you to enter your packet comments, so let's say, `This packet looks bad.` or whatever you might want to enter:

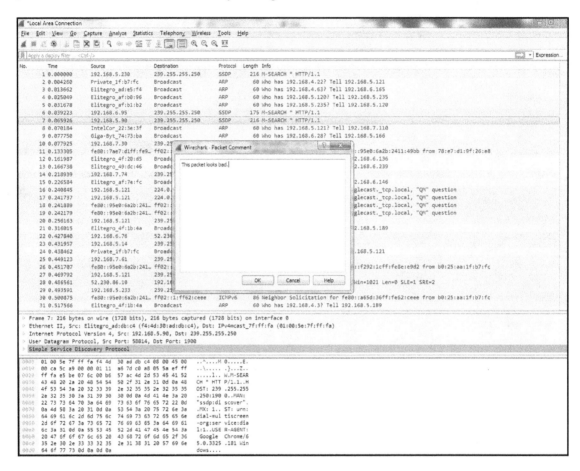

Click on **OK**. When you do so, you'll see that in the packet details area the **Packet comments** section pops in, and is nice and bright green so you can see that. And if you expand that, it'll actually show you the comment that you've entered:

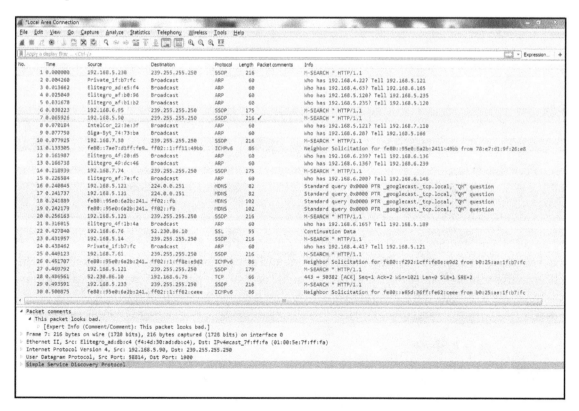

Now, if I select a different packet, it doesn't show up. So if I go back to the packet that I selected, then it'll display again. Now, you might be wondering: how do I find packets within a capture that have comments? There are multiple ways of doing this. One of them is to right-click on the **Packet comments** field in the details area here and we'll go to **Apply as Column**. When you click on that, it'll create a column, which will show you whether or not that packet has a comment on it:

Additionally, I can go to the expert info button that I was talking about in the bottom left-hand side. When you click on that, it gives us a whole bunch of information about our capture, which we'll ignore for now. But at the very bottom, there's the **Comment** section, and it will say, packet number **7** has a comment, as shown in the following screenshot:

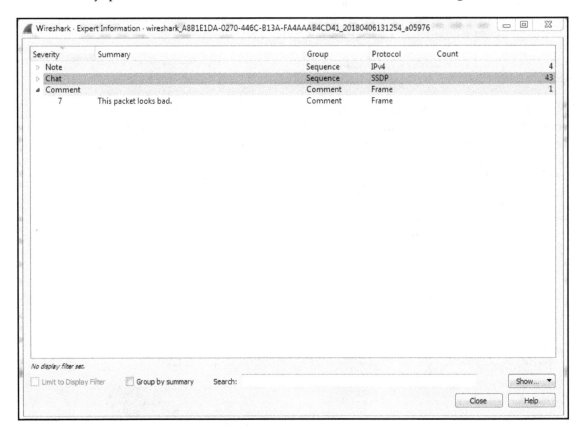

Now what's interesting is, if I move this to the side and we go select a different packet—and we'll go all the way to the bottom and choose packet number **60**—and then if I click on the comment in packet number **7**, you'll see that the packet list automatically jumps up to packet **7**, selects it for me, and shows me the comment. Isn't that nice?

A third way to find packet comments is to right-click on **Packet comments** and then go to **Apply as Filter** | **Selected**:

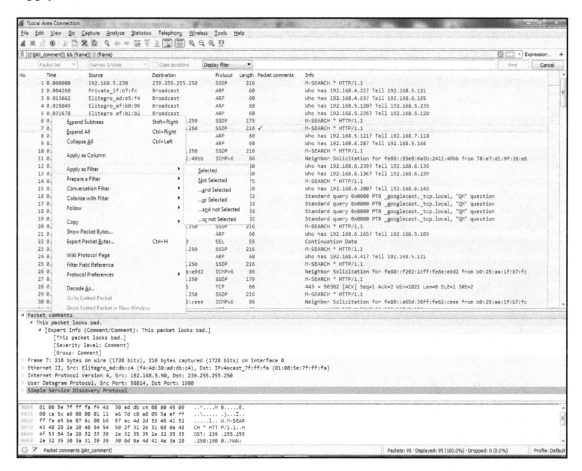

And when you do so, it'll filter your packet list by `pkt_comment`, and that's the field name for whether or not a packet has a comment in it. And you'll see here that packet number 7 is now the only packet showing because that's the only one we've made a comment for:

So, if I were to clear this, and we add another comment on another packets Comment 2, and if I reapply my filter on **Packet comments**, we can see that we have two packets there. So, that's another way of being able to find comments in them:

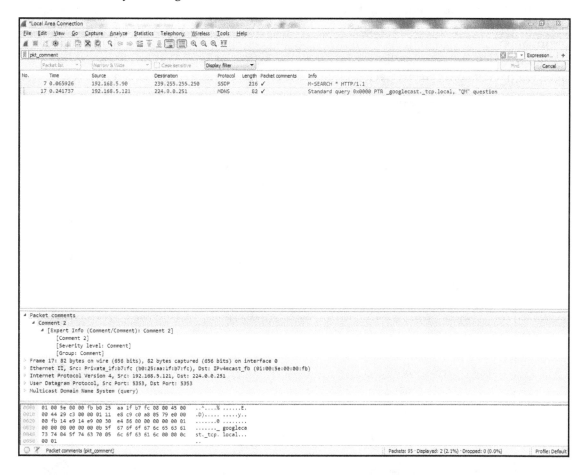

In order to print your capture or print an individual packet, you can go to **File | Print...** and you'll see a number of options here which look similar to the save and export dialog. So what we can do is print a **Summary line** for each packet, and if you uncheck this you will see it actually changes the preview as you go, so you can kind of see how the file's going to look. There's a summary line, which gives us information about each individual packet that's in the list, and the summary line looks kind of like the packet list view. So there's going to be one line, which is packet 1 and some information about it then another summary line for packet 2 and some information about it, and so on. So that's actually a handy one to have on.

Details: will show you the packet details list of the information about the different protocols, so we can turn that on or off. If I turn that off, that basically just shows us the packet list view. I'm going to leave that on. And then I could also include the **Bytes**, if I really needed to. You're not going to want to do that for a lot of packets. Obviously, your print would be very large, but you can see if I turn that on it'll show you the actual byte information, as shown in the bottom bytes. So I'll keep that off for now. And you notice in the bottom section here, just like we had with the export dialog, you can choose option **Selected packets only** or **All packets**. You can also select **Marked packets only**:

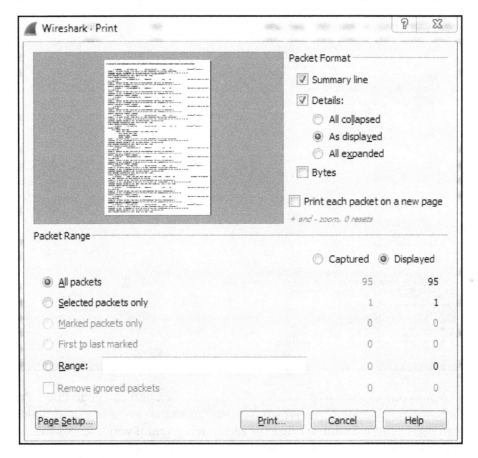

How to mark packets

What you can do is right-click on a packet and mark it, or do *Ctrl + M*. And you can mark a whole bunch of them, and they can be anywhere in the capture, it doesn't matter where-they don't have to be contiguous; and we'll mark up a bunch.

We'll go ahead and print this, and you can see the file that we printed here. I printed it as a PDF file so that it would be easy to show you. You can see that the summary line for each packet is displayed here. The summary line, which is basically the packet list line, includes the packet number we had when the packet came in, the source and destination, the protocol, and so on, along with the details that we wanted it to print. So that's the very basic information about which protocols were involved in the packet that it captured:

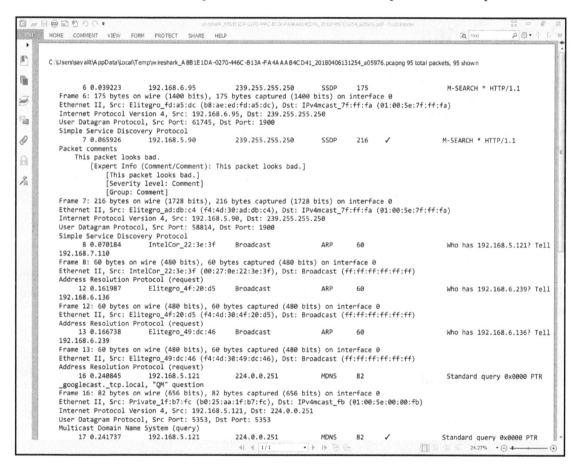

So, in this section, you learned how to create annotations and comments for your packet capture, as well as annotations or comments for individual packets, how to view them and find them in different locations in the interface such as on a column or a filter, or within the packet details view, expert information, and, additionally, how to print packets. Next up is remote capture setup. We'll go over how to capture your data from a remote machine from your Wireshark GUI, running `libpcap` on that remote machine.

Remote capture setup

In this section, we'll take a look at the prerequisites for using remote capture, specifically with WinPcap, which is a Windows port of the `libpcap` library and the configuration of remote packet capture on the remote device.

Prerequisites

The first prerequisite is to install the `pcap` libraries. I'm using a Windows computer here as our test machine, so I'll install the `WinPcap` libraries, which are a Windows port of the `libpcap` libraries originally written for Linux. So what I'll do is go to `https://www.winpcap.org/` and download the `WinPcap` libraries. Once it finishes downloading, I'll go ahead and just click on **Next** with the installer, and there's no need to customize anything there.

> The `WinPcap` libraries that are installed also come with Wireshark when you download Wireshark as a bundle. And so if you already have Wireshark installed on a system, most likely you already have `pcap` installed as well.

The next step is to set up a local administrator account which is going to be a service account for the `pcap` service, the remote `pcap` service that is running on this remote system. That is used in the authentication of the system that's running the Wireshark GUI when we add the remote interface. In order to do so, we'll perform the following steps:

1. Press the Windows key + *R*; it'll bring up the **Run** command.
2. Enter `control userpasswords2`.
3. Go ahead and click on **OK**.
4. We will go to **Advanced** and click on **Advanced** again.
5. Go to **Users** and we'll make a **New User...**; we will call this `pcap`.
6. We'll say it's a `Service Account` and give it a password:

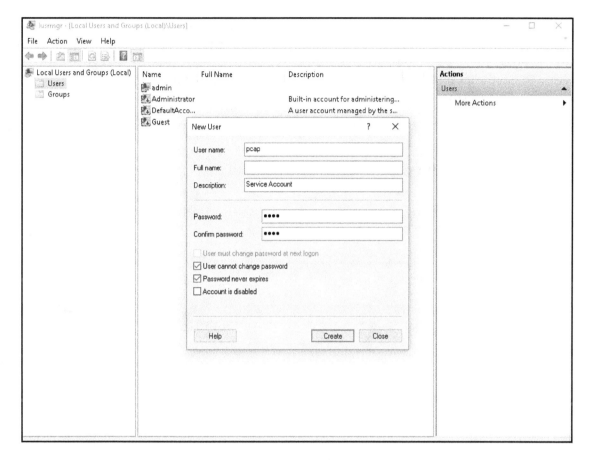

7. I'll uncheck the first option there to force the user to create a new password since this is a service account. We'll also prevent it from changing its password and never let it expire. We certainly don't want the password changing on the service account as we're trying to use it.

8. Now that we have our `pcap` service account, we need to give it administrator privileges. So we do that by right-clicking on **pcap**. Go to **Properties | Member Of |** and we're going to click **Add....** Type in `Administrators`, and go ahead and click on **Check Names**, which tells it to verify that the group Administrators is correctly typed in:

9. Click on **OK**, and we can go ahead and click on **OK** again.
10. Now the `pcap` service account has administrator privileges. The last step for setting up the remote `WinPcap` service is to press the Windows key again and then press *R*. We'll bring up our **Run** prompt and type in `services.msc`.
11. Go ahead and press *Enter* or click on **OK**, and we'll scroll down until we see **Remote Packet Capture Protocol v.0 (experimental)**. Although it says v.0 and experimental, this has been a service that has been available for a long time now, and I've never had any problems with it.
12. Go ahead and right-click on it. Go to **Properties** and click on the **Log On** tab.

13. We'll select the **This account:** radio option and **Browse...** to the `pcap` user, or whatever you named yours. Enter that and click on **Check Names**. It verifies that it's the correct spelling and it found the account.

14. And then we'll enter in the password that we gave it. All right. Click on **OK**:

It'll say it's been granted service rights; that's good. And at this point, we can tell it to start the service.

You can do so by clicking on the little play icon at the top; click on the **Start** shortcut there; or right-click on the service until it starts.

It should say running at this point. What I like to do, just in case, is to go up at the top here and click on refresh—just refresh a few times; make sure that the service didn't crash at all.

At this point, everything is set up and ready to be used. The last thing to check is to ensure that the Windows Firewall is either turned off, or port 2002 is enabled to pass through the Windows Firewall. For the sake of simplicity, we'll turn it off, so push the Windows key, bring up the Windows **Start** menu, and we'll search for the firewall. So in Windows 10 here, you can just type and it'll start searching. I'll go to the **Control Panel** in Windows Firewall, and you'll see here that it is currently enabled. So we'll click on **Turn Windows Firewall on or off**, and I'll turn it off. Click on **OK** and we can close that:

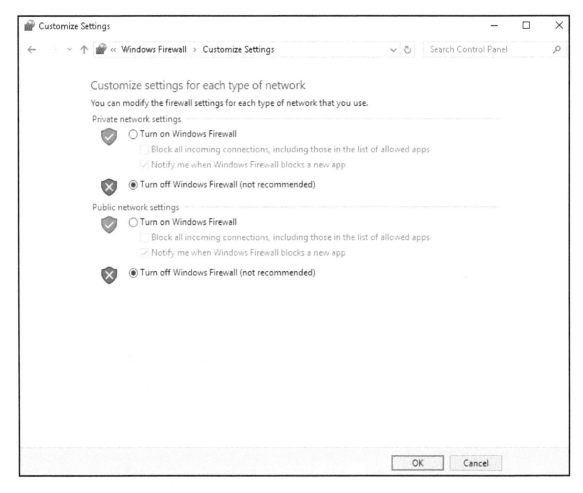

And that's all there is to it to set up a remote pcap system. So, in this section, we've gone over all of the configuration and installation needs for a remote system. Next up, we'll go over the remote capture usage, and how to set that up in a Wireshark GUI and start capturing traffic from a remote device.

Remote capture usage

In this section, we'll take a look at how to use that remote packet capture software that we set up with WinPcap on the remote system.

In order to use that remote WinPcap service running on the remote system and capture packets from it, we need to add that into our local Wireshark interface so that we can capture it. So in order to do this, we will perform the following steps:

1. We will go ahead and click on **Capture options** icon.
2. Click on **Manage Interfaces...** and you'll see here that there's the **Remote Interfaces** tab; click on that.
3. Click on the plus icon in the bottom left-hand side here.
4. Enter in the **Host** IP address of that remote system.
5. Click on the **Password authentication** radio button, and enter in the credentials for that service account that we created. I used `pcap` here. You can then enter in the username and password and click on **OK**. At this point, it should show us the remote interfaces that it sees on the other device. So you see here that's my `5.25` device, and here's the interface that it detected:

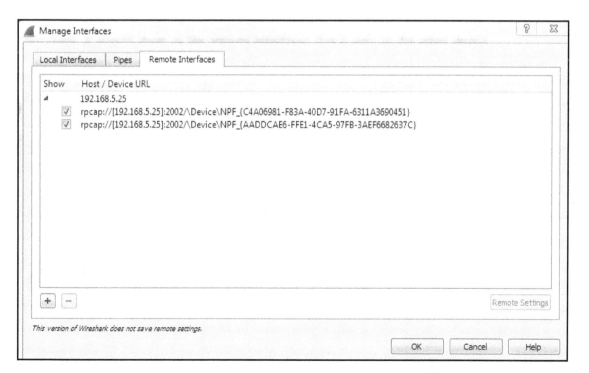

If you do not see this at this point, or you get a popup saying that you have some sort of connection error or it can't connect to the remote host, or anything like that, make sure that the service is running. Remember, when we set up the service on the remote system, it was on manual for the service—it was not automatic. So there's a good chance that the server's stopped or the system has rebooted, or something like that. Go over there and make sure that the service is enabled.

6. So we go ahead and click on **OK**. You'll see that it shows in our interface list here. We can then go ahead and click on **Start**:

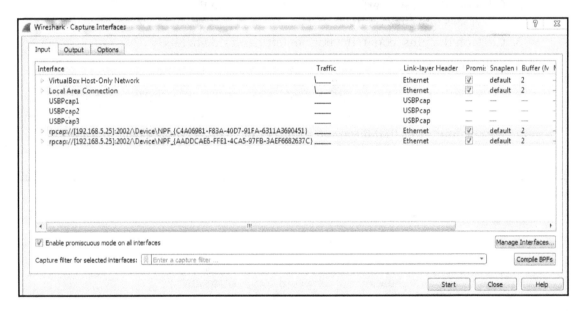

And that's all there is to it.

Summary

In this chapter, you've learned a number of skills in Wireshark, including what's new in Wireshark 2, and there are some features from 1.8 that I mentioned, namely, how to capture traffic on a local Wireshark installation; how to get the packets to your Wireshark installation through different means, such as SPAN ports; saving and exporting these packets in different ways; annotating or adding comments to the packet captures and individual packets, and printing them, or selections of them; and lastly, setting up the remote packet capture with WinPcap on a remote Windows system, and using that packet capture from the remote system to a local Wireshark installation.

In Chapter 3, *Filtering Traffic*, we'll go over the various ways of filtering traffic, both with the capture filters, as well as display filters, and also look at additional ways of filtering things.

3
Filtering Traffic

In this chapter, we'll cover the following topics:

- **Berkeley Packet Filter (BPF)** syntax
- Capturing filters
- Displaying filters
- Following streams
- Advanced filtering

Berkeley Packet Filter (BPF) syntax

In this section, we'll take a look at the BPF, its syntax, and some of its history.

So, let's talk about BPF's history. Many years ago, every operating system had its own packet filtering API. There are a number of examples, such as Sun, DEC, SGI, and Xerox. They all had their own operating systems, and each operating system had its own API for capturing and filtering packets. So, when you needed to do network analysis, you had to use their specific software, which is built into the operating system, and their specific filtering capabilities within the API that they designed. That made it very difficult because depending on the implementation of your network and what different operating systems were involved, you had to know all these different APIs and all of these different filtering rules in order to get anything done. So, in 1993, Steven McCanne and Van Jacobson released a paper titled *The BSD Packet Filter (BPF)* and they outlined the rules and the ideas behind BPF and explained how it could be a standardized method for filtering the captured traffic. It just so happened that it caught on and became very popular, especially as `libpcap`, WinPcap, and other libraries out there began to utilize BPF as its standardized filtering system, and especially with the use of Wireshark nowadays which utilizes these libraries.

In order to write BPF, you need to create an expression, which contains one or more primitives, including an ID, such as a name or number, an IP address, or an Ethernet address plus a qualifier. A qualifier has the following three pieces to it:

- Type
- Direction
- Protocol

For a type, it could be an individual host, a network, a port, or a port range. The direction can be either the source or the destination, or the source and destination. And the protocol is either Ethernet, FDDI, Wireless LAN, IP, IPv6 nowadays, ARP, RARP, DECNET, TCP, or UDP. You need to define these different pieces that you want, and how you want to limit your traffic and the values that go with them—the ID, the name, or number to go along with these qualifiers. So, I have some examples for you so that it can make some sense as to how to create a BPF expression. The first one is `ip host 192.168.1.1`. The IP is the protocol, the host is the type, and the ID is the IP address. This will filter the traffic for that IP address, whether it's the source or destination. This host keyword does both of these for us.

Next up, I have `ether src AA:BB:CC:DD:EE:FF` and a fictitious MAC address. This has the same idea as the IP host. We're defining Ethernet as our protocol, the source as the direction, and the MAC address that we're looking for.

If you happen to be capturing traffic that has multiple VLANs, such as spanning a port that's a trunk port on a switch, you can specify the VLAN(s), for example, `vlan 100`.

The next example is `ether broadcast`, and this one has a special keyword being used for broadcast to tell the BPF that we want to filter all of our traffic, if it's a broadcast of some kind, on layer 2.

And my last example is `tcp port 80`. So, we'll filter that traffic looking for HTTP traffic most likely—looking for only port `80` of any source destination.

Up next is capture filters, where we'll take this BPF syntax and apply it to interfaces within Wireshark.

Capturing filters

In this section, we'll take a look at how to filter traffic before it's captured with the BPF syntax. So, we'll filter that traffic on the capture interface.

In Wireshark, there are two places to enter a capture filter.

The first one is right on the following main screen. Right in the middle, we have the capture section, and it says, **...using this filter: Enter a capture filter**. So, we can actually do that on the main screen. Try to enter a capture filter, then it will start capturing with that applied filter. You'll also see that there's a green bookmark icon, as shown in the following screenshot. If you hover over that icon, it says **Manage saved bookmarks**. And if we click on that, there's a number of saved bookmarks that are already built into Wireshark. So, if there's a common function that you want to filter on, it may already be in the list:

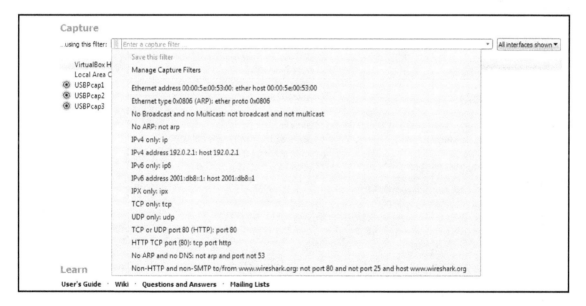

But you can also go up to the top and click on **Manage Capture Filters**. This gives you a list of all of your predefined capture filters and any that you have saved yourself, so you don't have to keep entering the same capture filter over and over again. You can create one and save it. So, all you need to do is click on that plus icon, and then you can enter in whatever it is you want to do. So we could do `ip host 192.168.5.25`, and we've now created a new capture filter. Then, of course I can rename it. If I double-click on that, it will allow me, to rename and I can say `My Host`:

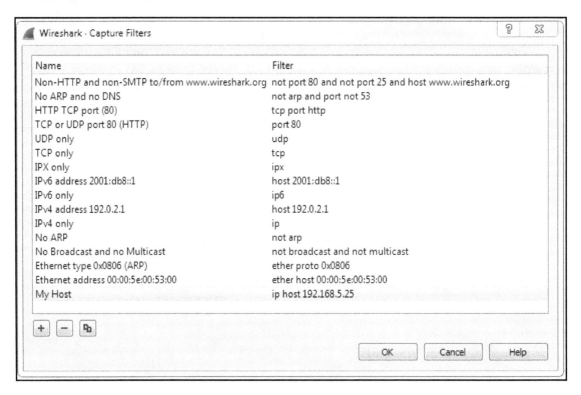

Now, if I go back and look, it would be visible now.

You can also save a capture filter by selecting an interface first and then entering what you wish. You'll see that it turns green if it's a valid entry. Then, click on that bookmark icon and then on **Save this filter**:

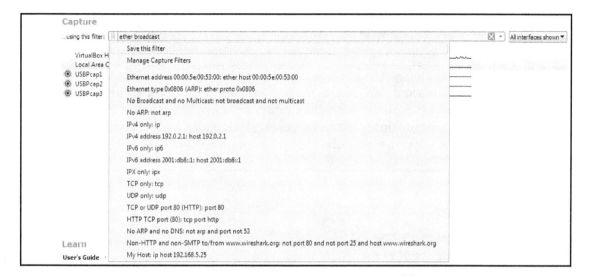

If you click on that, it'll then bring up the **Manage Capture Filters** window prepopulated so that you can simply name it as you wish. Then, using this capture, I can simply double-click on my interface and begin capturing that traffic. You will see that I'm filtering on Ethernet broadcast, so it's only going to show the broadcast traffic on layer 2:

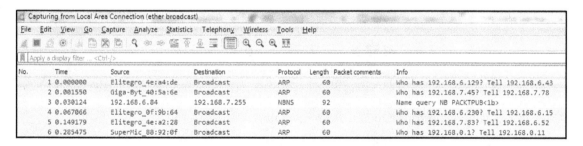

The second way of entering a capture filter is through the gear icon:

Click on **Capture options** and you'll see at the bottom that there's the **Capture filter for selected interfaces** option. You can enter the capture filter just like you did in the previous window, as well as manage your bookmarks. So, we could enter in `ip host 192.168.5.25`. Now, select the interface, then it'll turn green, and I can start my capture:

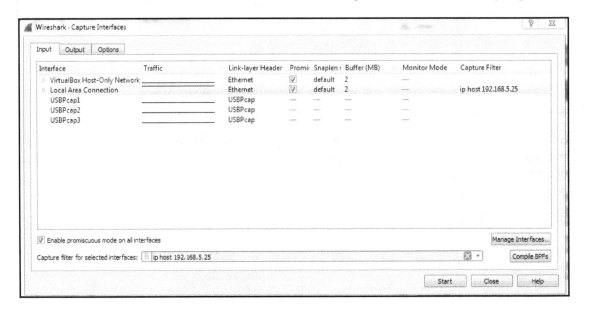

Now, it'll show me any traffic to and from my gateway.

Our next section is *Displaying filters*, where we'll filter our traffic after the capture has already been started or it's already been saved, rather than filter the traffic during the capture itself. This is a very common method of doing things so that you capture all the traffic on your network, and then simply view only the pieces you need.

Displaying filters

In this section, we'll go through display filters. And, in regards to display filters, we'll talk about how to sift through a large packet capture. So, a common method of capturing traffic is to not use the capture filter and instead capture everything that the interface can see, and then simply filter out exactly what you want to view because some of these other packets may be useful in diagnosing some sort of problem. We'll also go over quick access filter buttons. So, if there's a common thing that you need to filter on all the time in your environment, you can create a quick access button and simply click on that, and you don't have to type out the display filter every time. And there are a number of filter operators that you can use to combine multiple filters to create a full expression as to what you want to specifically filter on:

- `eq/==`
- `le/<=`
- `or/||`
- `not/!`
- `and/&&`
- `ne/!=`
- `gt/>`
- `contains`
- `lt/<`
- `matches`
- `ge/>=`
- `()`

So you could combine the IP address filter with a port number filter, or something related to TCP, or something related to two MAC addresses. You can combine them or exclude them in different ways using these filter operators. You can use either the Word version of the operator, which is what's before that /, or you can use whatever's listed after the /, which is like the mathematical equivalent of the Word option. There's also a parentheses, so you can also combine filters and their operators together, kind of like a mathematical equation. So, you can have certain comparisons, such as two OR statements, compared first, and then have it with something else as an addition, such as an AND statement.

So, to work with display filters, let's get some traffic first. I'll start a capture on my primary interface without any capture filters. So, we will have some packets coming in, which will be whatever's idle on my system at the moment.

You'll see up at the top of the screen that we have a **Apply a display filter** entry box. And it also has that bookmarks icon, just like the capture filters did:

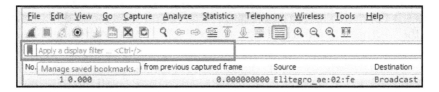

So, let's get some packets to work with. I'll start a capture on my primary interface, without any capture filter entered. So we'll capture everything that's occurring on my system; currently, whatever's in the background. So, after we have a good amount of traffic to work with, you will notice that at the top we have a textbox that says **Apply a display filter**, and that works just like the capture filter textbox does. We can type in whatever is the display filter that we're looking for. We also have the bookmarks icon on the left, just like with the capture filters. And if I click on that, it gives me my saved filters. I can have a whole bunch of them listed, as shown in the following screenshot—these are the common ones that come prebuilt with Wireshark:

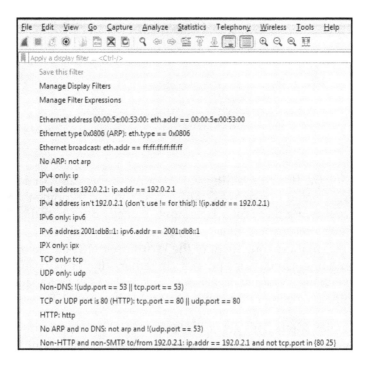

Just like with the capture filter, if I enter one filter into the textbox, I can save it. Alternatively, I can go in and manage my display filters, and I can add and remove them manually, as well, just like with the capture filters.

Let's start off with filtering by something. So, let's do `ip.host`, which is `eq` `192.168.5.25`, so that'll be my gateway:

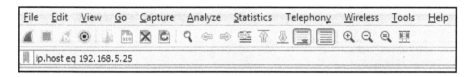

Then, I can either click on the arrow on the right-hand side to apply it or press *Enter*. And when I do so, there's all the traffic to and from my gateway `5.25`:

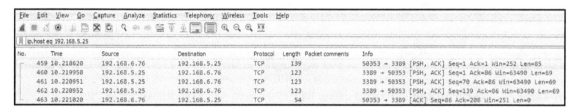

We can also filter by source or destination. We will use the keyword `host`, so we will input `ip.host`.

Host is a shortcut that could be included both at the source and destination traffic for a specific IP address.

We could also filter by source, which is `src`, but if we do it this way we'll only get the traffic originating from `5.25`; we'll not get the return traffic. If we do the same thing with destination, `dst`, we'll get the return traffic, but not the originating traffic.

So, how do we get this traffic? We can chain it together using one of the operators, like I mentioned. So, we'll tell Wireshark we want both these destinations in the source traffic. We have `ip.dst` up there already. So, what we'll do is combine it with an OR. And if you remember, you can either type `or`, or the two pipe symbols. And we'll type `ip.src == ` and put in the IP address of my gateway.

And now that it's a valid display filter, it's turned green. I can just press *Enter* and there we go:

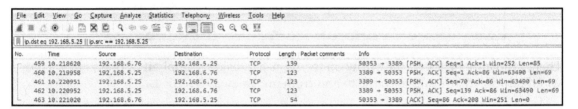

Now we have both directions of traffic, just like if I'd typed in `ip.host`. So you can see how using some of these shortcuts that you know of in the display filter options can make things faster for you.

Another interesting thing we could do is, let's say we need to troubleshoot something with a slow transfer with TCP. So we want to look at the window size, so we type in `tcp.window_size_value < 10`. So, let me press *Enter* and there we go. So, we have one packet. Remember, we didn't define any sort of IP address. Any of the packets that Wireshark has captured that has a window size less than `10` will be shown to me:

If we expand our TCP details, you'll see there's the window size; it's certainly less than `10`:

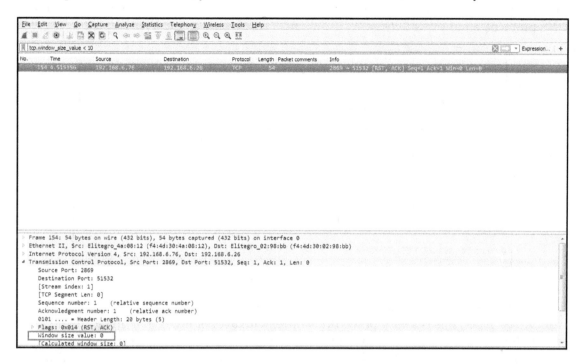

So, let's create an example using the parentheses I mentioned. So, what we'll do is filter based on my gateway again. What I'll do is quick-access one of my last used display filters. So, I'll type `ip`. Now, if you look there, it has the most recently used ones at the top. So, I'll use the down arrow, select the one I want to use, and press *Enter*:

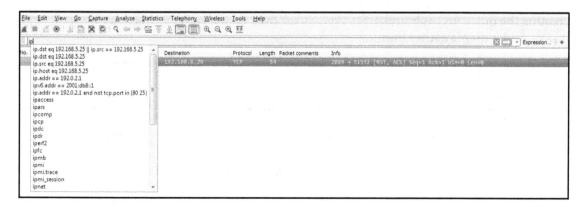

We'll add some parentheses to that. Now, I'll add in `dns`. So I want to see any traffic to or from my gateway, and it's a DNS entry. So I'll type `&& (dns)`, which is a shortcut for the DNS protocol, and I press *Enter*:

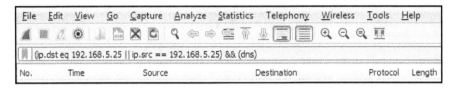

Now, we can go even farther. So we can say `&& ip.len <= 72`. And now, I've trimmed it down to any DNS packets to and from my gateway that are less than 72 bytes in length:

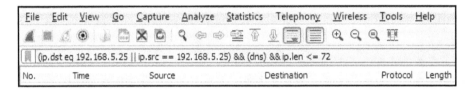

So, you can see how this is really powerful, and how you can trim off exactly what you want to see. Now, this is a very powerful feature, with a lot of filters in it. In fact, there are tens of thousands of filters for two thousand protocols. So, if you go to `https://www.wireshark.org/docs/dfref/` for display filter reference, it will show you all of the different fields that you can use. So what we'll do is we'll click on **I** and look for **ip**. There we go. Let me click on that, and then that will tell us all of the subfields that we can filter on. So, `ip` is the protocol. Now, we're curious about, maybe, the address or the checksum, destination, source, types of flags, GeoIP information, and so on. So this is a great reference to search for what you wish to filter by. Now, the last feature we'll talk about for display features is creating the quick access buttons. So, if you look over on the right, we have a plus icon, and it says **Add a display filter** button:

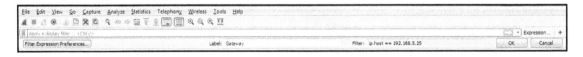

So, if we click on that, we can then create a **Label** and enter a filter to save. So, let's say I want to make one for my `Gateway`. I'll make a **Filter** by saying `ip.host == my gateway`. Once you click on **OK**, you will see a button for **Gateway** now. So if I have a big capture with hundreds and thousands of packets, I can simply click on my **Gateway** button and it instantly applies that filter:

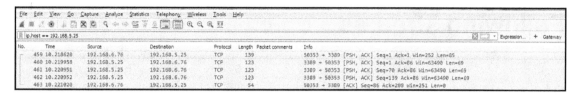

Now, if you wish to remove some of these buttons, the easiest way to get there is to simply click on the plus icon again, and then there's a button on the left which says **Filter Expression Preferences....** That will take you directly into Wireshark's **Preferences** section for this, and then you can just edit or remove whatever buttons you need to change or you don't want:

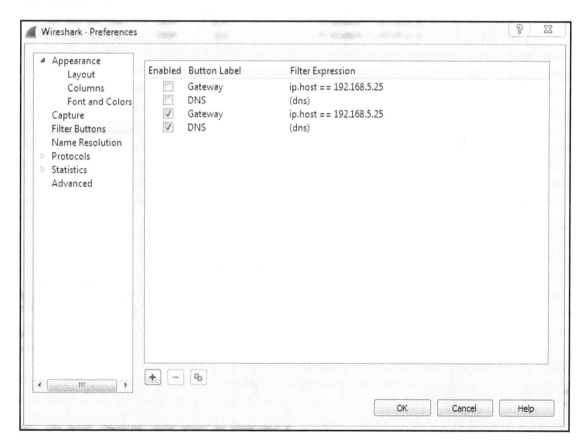

So, in this section, we went over the basics of the display filter and created quick access filter buttons. Up next, we'll talk about filtering by TCP and UDP streams and show specific conversations within our packet captures.

Following streams

In the previous section, we went over how to use display filters to limit what you see in a packet capture. In this section, we'll build on that and follow streams. What that means is, in this section, we'll follow TCP and UDP streams to pick out conversations within our packet capture so that we can view the specific communications between each TCP communication or each UDP communication. So, what we have is a packet capture of opening up the `https://www.cisco.com/` home page. And the `https://www.cisco.com/` home page is not encrypted with SSL by default, so we can see all of the HTTP communications within it without having to add in some sort of SSL key to decrypt it:

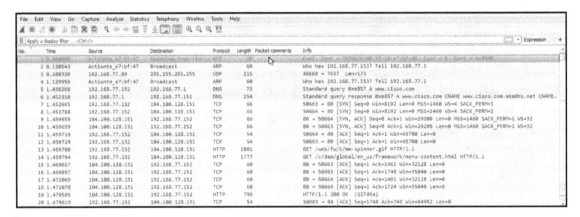

In the beginning, we can see the DNS query to Cisco and the response, and then the beginnings of the TCP handshake. Then, we start to retrieve some files and some HTTP traffic for retrieving HTML. Now, we could of course apply what we learned in the last section, and go up to the top and create a display filter for the web server as the host and the TCP protocol, with some other parameters to try and filter out what we want to see, but we'll still have a whole lot of data. You can see this is a very large capture, just to load up one web page.

So, how can we pick out individual communications within this packet capture? Because one thing you'll note is, in a web page, you have multiple files, graphics, CSS files, JavaScript, or whatever it might be that it needs to retrieve. There are a number of different files it has to pull in order to build the web page that you see. Each one of these is its own communication and its own TCP stream. So, what we need to do to follow a TCP or UDP stream is to select a packet within the capture that is within the stream that we wish to view. So, for each of these individual files in this web page, each one's going to be its own TCP stream. So we need to select a packet within that, and then follow the stream.

So, let's choose one of the graphics. Let's go up to our spinner.gif. As shown in the following screenshot, we'll right-click on our HTTP packet and go to **Follow | TCP Stream**:

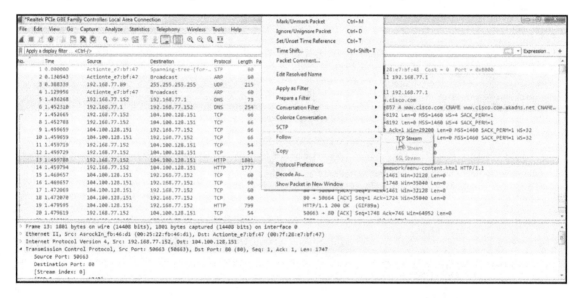

By default, the data shows up in an ASCII format, and that's usually very helpful to you because a lot of the traffic you're probably looking at is text-based. But you may wish to change that when you're doing this follow stream. You can do that at the bottom, where it says **Show data as ASCII**. You can change that drop-down box and select whatever data format you wish. We'll leave it as **ASCII**, since we're looking at some HTTP traffic with a GIF transmitted within it. And you can see here that we have red and blue lines. Red is the client and blue is the server, and you can see that at the bottom, where it says **16 client pkt(s)**, **228 server pkts(s)**:

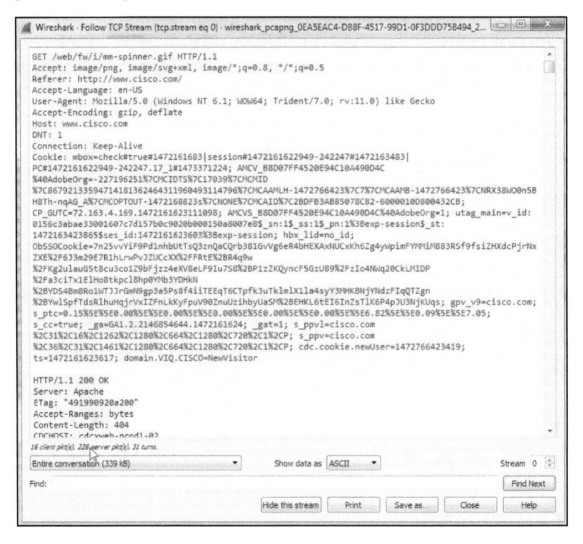

We can also change this drop-down box and select the communication that we wish to look at. And as we scroll down in this data view, you will see the blue and red of the client and the server sending their packets back and forth. And, as you may have noted in one of the earlier sections, if you click on the data, it will jump to the packet in the packet view. An additional feature of following TCP is that you can **Print** or **Save as...** your packets. So you can actually export these and save them as whatever file format you deem.

You can also follow UDP streams.

 UDP streams are more difficult to follow, though, so it may not always work perfectly, but Wireshark does the best that it can. The difference between TCP and UDP is that TCP is reliable and UDP is unreliable communication.

Now, you can do the same thing as with following TCP traffic, that is, right-click on a packet and go to **Follow**, and you can see that **UDP Stream** is available for us to click on.

Select **UDP Stream** and it'll do the exact same thing. It will show the data up in the top pane. It may be useful or not, depending on what kind of data it is. It will allow us to perform all the same features as following the TCP stream.

In this section, you learned how to follow TCP and UDP streams and to pick out specific conversations within your packet capture and data flows. Next up, we have advanced filtering, where we'll get into picking out more nuanced sections within a packet capture and within your packets to enhance your ability to filter your traffic.

Advanced filtering

We'll go over how to filter traffic in data fields within packets, how to create columns on specific fields in our data packets and to sort them, and filter with these more hidden methods that you may not have noticed yet.

We'll go to our packet capture, opening up `https://www.cisco.com/`. And what we'll do is expand the section that gives us the details of our packet. We'll scroll down and find a packet that we wish to investigate a bit more.

What we can do is expand the sections, which will allow us to easily view the different fields of data within the different layers of the packet. So, if you're familiar with the OSI layers that we use in networking, then this'll look familiar to you. So, we have our layer 2 information with our frame and Ethernet, layer 3 with IPv4, layer 4 with TCP, and so on.

So, let's click on TCP to expand this, and we'll go down to **Window size value**. Now, the window size is an important field that we'll get into in more detail later on, but what we'll do is view this as a column. So, let's right-click on **Window size value** and click on **Apply as Column**.

What that will do is add a new column in our packet list, with the window size value for each packet. Now, you'll see it right-aligned, and with that right up against the information it's a bit hard to see. So, let's right-click on that header up there and go to **Align Center** and click on it:

No.	Time	Source	Destination	Protocol	Length	Packet comments	Window size value	Info
580	18.226733	192.168.6.202	192.168.7.255	BROWSER	243		Align Left	ment PPMUMCPU0221, Workstation, Server, NT Workstation, Po
581	18.227306	Elitegro_90:2e:f1	Broadcast	ARP	60		Align Center	168.6.130? Tell 192.168.6.137
582	18.231108	Elitegro_01:9f:b4	Broadcast	ARP	60		Align Right	168.6.137? Tell 192.168.6.130
583	18.235852	Elitegro_4a:0c:07	Broadcast	ARP	60			168.6.250? Tell 192.168.6.93
584	18.252076	192.168.7.120	239.255.255.250	SSDP	216		Column Preferences...	TTP/1.1
585	18.260175	192.168.6.112	192.168.6.76	TCP	66		Edit Column	[SYN] Seq=0 Win=64240 Len=0 MSS=1460 WS=256 SACK_PERM=1
586	18.260353	192.168.6.76	192.168.6.112	TCP	66		Resize To Contents	[SYN, ACK] Seq=0 Ack=1 Win=8192 Len=0 MSS=1460 WS=256 SACK
587	18.260844	192.168.6.112	192.168.6.76	TCP	60		Resolve Names	[ACK] Seq=1 Ack=1 Win=525568 Len=0
588	18.261216	192.168.6.112	192.168.6.76	HTTP	290		✓ No.	t/udhisapi.dll?content=uuid:1872da05-f057-4ba1-bb74-9eaf54d0
589	18.262662	192.168.6.76	192.168.6.112	TCP	259		✓ Time	[PSH, ACK] Seq=1 Ack=237 Win=65536 Len=205 [TCP segment of
590	18.262826	192.168.6.76	192.168.6.112	HTTP/X...	2450		✓ Source	OK
591	18.263726	192.168.6.112	192.168.6.76	TCP	60		✓ Destination	[ACK] Seq=237 Ack=2602 Win=525568 Len=0
592	18.317287	Elitegro_af:b1:e7	Broadcast	ARP	60		✓ Protocol	168.7.16? Tell 192.168.6.116
593	18.319068	Elitegro_af:b1:e7	Broadcast	ARP	60		✓ Length	168.5.18? Tell 192.168.6.116
594	18.327017	23.102.45.176	192.168.6.76	TCP	66		✓ Packet comments	[SYN, ACK] Seq=0 Ack=1 Win=8192 Len=0 MSS=1460 WS=256 SACK
595	18.327221	192.168.6.76	23.102.45.176	TCP	54		✓ Window size value	[ACK] Seq=1 Ack=1 Win=66048 Len=0
596	18.355976	23.102.45.176	192.168.6.76	TCP	1514		✓ Info	[ACK] Seq=1 Ack=219 Win=131328 Len=1460 [TCP segment of a r
597	18.355981	23.102.45.176	192.168.6.76	TCP	1514			[ACK] Seq=1461 Ack=219 Win=131328 Len=1460 [TCP segment of
598	18.356618	192.168.6.76	23.102.45.176	TCP	54			[ACK] Seq=219 Ack=2921 Win=66048 Len=0
599	18.356092	23.102.45.176	192.168.6.76	TCP	1514		Remove This Column	[ACK] Seq=2921 Ack=219 Win=131328 Len=1460 [TCP segment of
600	18.356093	23.102.45.176	192.168.6.76	TLSv1.2	1199			, Certificate, Certificate Status, Server Key Exchange, Ser
601	18.357186	192.168.6.76	23.102.45.176	TCP	54			[ACK] Seq=219 Ack=5526 Win=66048 Len=0
602	18.374097	192.168.6.76	23.102.45.176	TLSv1.2	236			xchange, Change Cipher Spec, Encrypted Handshake Message
603	18.385799	192.168.4.69	239.255.255.100	IGMPv1	106			eport
604	18.388044	192.168.5.75	239.255.255.250	SSDP	216			M-SEARCH * HTTP/1.1
605	18.394374	fe80::3664:a9ff:fe6..	ff02::1:ff26:9ba8	ICMPv6	86			Neighbor Solicitation for fe80::1422:91de:4026:9ba8 from 34:64:a9:6e:5a
606	18.398262	Elitegro_af:b0:99	Broadcast	ARP	60			Who has 192.168.6.116? Tell 192.168.6.104
607	18.418599	192.168.6.76	52.114.32.7	TCP	66		8192	59697 → 443 [SYN] Seq=0 Win=8192 Len=0 MSS=1460 WS=256 SACK_PERM=1
608	18.427024	192.168.6.91	239.255.255.250	SSDP	216			M-SEARCH * HTTP/1.1
609	18.442104	192.168.6.30	239.255.255.250	SSDP	216			M-SEARCH * HTTP/1.1
610	18.504839	192.168.6.116	192.168.7.255	BROWSER	216			Get Backup List Request
611	18.505199	192.168.6.116	192.168.7.255	NBNS	92			Name query NB WK1b>
612	18.505586	Giga-Byt_7f:3b:39	Broadcast	ARP	60			Who has 192.168.6.116? Tell 192.168.7.179

```
[Stream index: 12]
[TCP Segment Len: 236]
Sequence number: 1    (relative sequence number)
[Next sequence number: 237    (relative sequence number)]
Acknowledgment number: 1    (relative ack number)
0101 .... = Header Length: 20 bytes (5)
> Flags: 0x018 (PSH, ACK)
Window size value: 2053
[Calculated window size: 525568]
[Window size scaling factor: 256]
```

 Click on that header and it will sort our packets, from smallest to largest or largest to smallest. Now, this is very useful because, many times with transfer issues, you'll have a window size problem. So, it might be beneficial to you to sort this way and look for any window sizes that are, very small, depending on the packet. Not every packet that's small in its window size is a bad thing, but it's something useful that you could take a look at. Now, you can do that for almost anything; we can go in and make columns for almost anything.

Additionally, we can remove the column. We'll right-click and go to **Remove This Column** to remove it.

We can go into TCP (or any of these fields) and create filters based on what we see. So, it's much easier than going into the display filter field and trying to find exactly what you want to do because, as I mentioned before when I showed you the website, there are hundreds of thousands of these different things that you can look for. So, instead of doing it that way, you can do it visually with these packet details, and select what you want to filter on. So, let's filter on **Source Port**. We will right-click on **Source Port** and go to **Prepare a Filter | Selected**:

What that will do is prepare a filter in the top section, with the source port information selected. So click on that, and it has `tcp.srcport == 50031` (that's the shorthand for source port):

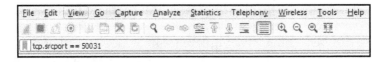

Now, if we apply this, it'll only show source ports that are exactly `50031`. What if we want to change that? I can of course go back and change the source ports, as you saw before in the operators and such. And then, we can go ahead and apply that. So now, I have all the packets listed that have a source port less than or equal to `50031`.

Let's find another one. Let's go to TCP again. We will scroll down, look into the **Flags**, and do checksum. So, let's right-click on **Acknowledgment**, and we'll go up to the top to **Apply as Filter** and click on **Selected**:

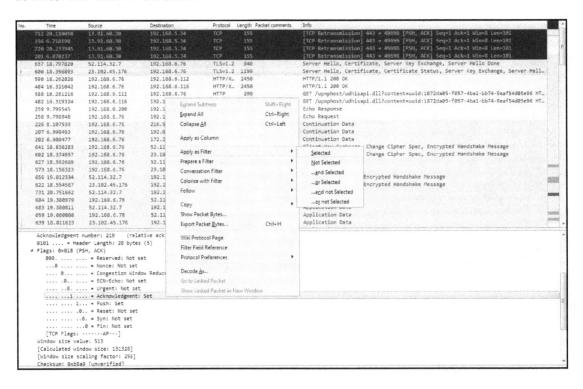

The difference between **Prepare as Filter** and **Apply as Filter** is that the former puts the filter into that top field—the display filter field-but doesn't apply it, while the latter does both. So, if you know exactly what you want, you can just go straight to the **Apply as Filter** option.

So, we now have all of the TCP acknowledgment packets listed. Now, we can also expand that, and let's go down and find another one—let's find another **Flag**. So, there's an **Acknowledgment**, and we'll add **Push**. So we want to see all the packets that are acknowledgments, but are also **Push**. We'll right-click on **Push** and go to **Apply as Filter** | **...and Selected**. You can see that it applied all of the syntax that's required to make this work. So, it takes a lot of the heavy lifting out of creating filters. So now we're looking at all of our packets that are acknowledgments with the push field set.

Let's look at another feature. What we will do is, we'll create a filter for my gateway again. So, there's `ip.host == my gateway`, and we apply it; now, that's all the traffic to and from my gateway.

Let's sort it by number so that it makes sense. What if I don't want to see the DNS? What I can do is go to DNS, right-click on it and go to **Apply as Filter** | **...and not Selected**.

And what that will do is negate the selection. So I select that, and you can see in the syntax it returns `&& !(dns)`:

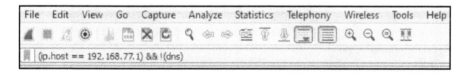

So any time you put the exclamation point in front of something, that tells Wireshark that you do not want to see that.

So, I highly recommend you spend time going through these packet details, learning where they are, what you want to look for, getting used to using **Apply as Filter** and **Prepare a Filter**, and understanding how they work together. You can actually take a packet capture that has hundreds and thousands of packets in it, and trim it down to just a few dozen that you actually care about.

Summary

In this chapter, we've learned about BPF syntax and its history and how to create BPF syntax. We also saw how to use that BPF and then apply it as a capture filter and reduce the packets that we end up capturing on our capturing interface. We then saw how to create and use display filters to prune what we have in a packet capture to what we just need to see. Furthermore, we saw how to follow streams, both TCP and UDP streams, so that we can view specific conversations within a packet capture and export that data if required. We also saw how to go into the different packet fields and lengths and all the different pieces of data within the headers of the packets and be able to create filters based on them.

In Chapter 4, *Customizing Wireshark*, we'll start tweaking Wireshark and actually customizing it and creating our own preferences and profiles so that we can make Wireshark our own.

4
Customizing Wireshark

In this chapter, we'll cover the following topics:

- Preferences
- Profiles
- Colorizing traffic

Preferences

To access Wireshark's preferences, go to **Edit | Preferences...**; this will open up the **Preferences** window. On the left, you'll see that there are a number of categories that you can choose from:

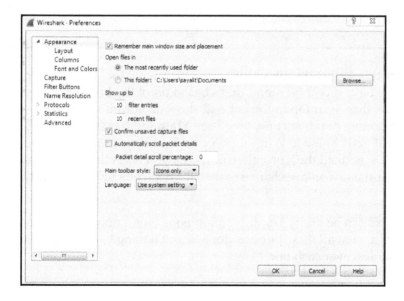

Appearance

The first category is **Appearance**, where you can change a number of settings, including the default folder that you most commonly open files from and the **filter entries** and **recent files** values. The **filter entries** changes the number of filters that appear in the drop-down box at the display filter section. So, right now there's 10, which you'll see once we close this:

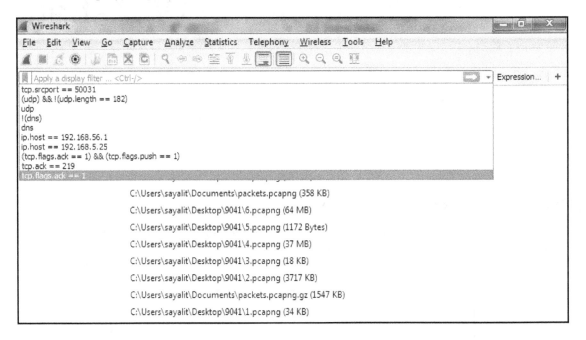

You can change this so that it shows more, and that's what that preference does. Additionally, we have recent files, and that's based off of the **File** menu. If you change that to a higher value, then your **Open Recent** will show an additional number of recent files. Down near the bottom of this section, you'll see **Main toolbar style**, and it says **Icons only** right now. And if you're new to Wireshark, you might have noticed that there is a toolbar up at the top there with all the icons; they don't tell you what they are unless you move your mouse over them. You can change that so it says **Icons only**, **Text only**, or **Icons & Text**.

This can be pretty helpful for someone who's new to Wireshark. It'll tell you what all these buttons do, without having to spend time to move your mouse over each one.

Layout

One of the helpful sections in here is the **Layout** section, underneath **Appearance**:

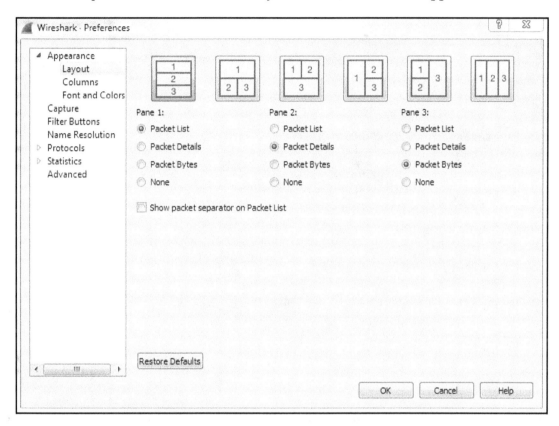

In the **Layout** section, you can change what that default view looks like within Wireshark. You might have noticed that, as I've done captures previously in this book so far, that the **Layout** section breaks the default view up in Wireshark into three panes, and the three panes are on top of each other. The top one being the list of packets, the middle one being the details of a selected packet, and the bottom one being the bytes. If you don't like that and you want to change how it looks, either the overall organization or removing one of these panes, you can do that here; you can see a number of options across the top as to how you might want things broken up. You could also change which information goes in which pane, and whether or not you even want any to show up, as some people don't really care about the packet bytes, but only about the list and the details. You can certainly go and turn off the **Packet Bytes** if that's not something that you need.

Columns

Now, in Chapter 3, *Filtering Traffic*, we went through how to create columns in Wireshark, and this is another way of doing so and editing these columns:

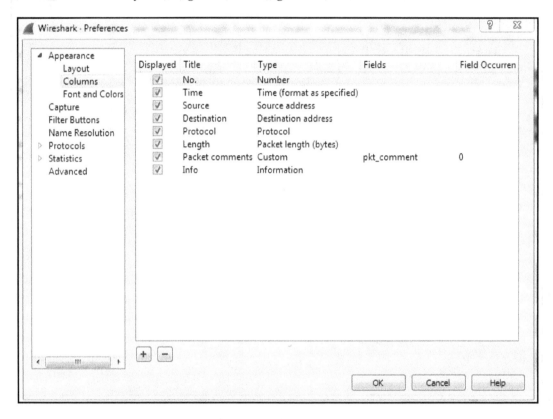

So, while you can right-click on a field within the packet details to create a column out of it, you can also create your own custom ones in here, or reorder or remove the different ones that are already there.

Fonts and colors

You can, of course, customize the fonts and colors within Wireshark, but these are separate from the colorizing rules that we will get into later on in this section:

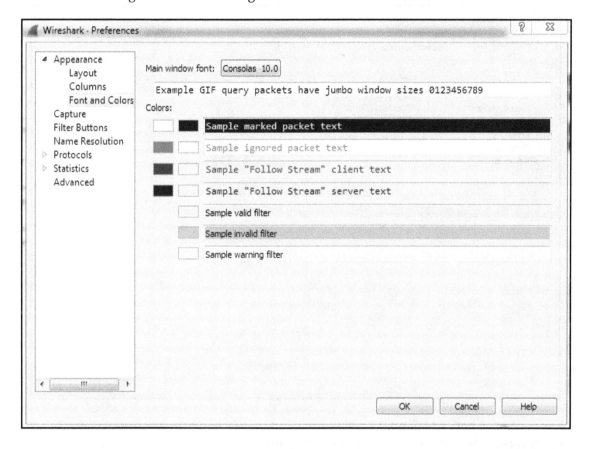

Capture

Under the **Capture** category, you can choose a **Default interface**:

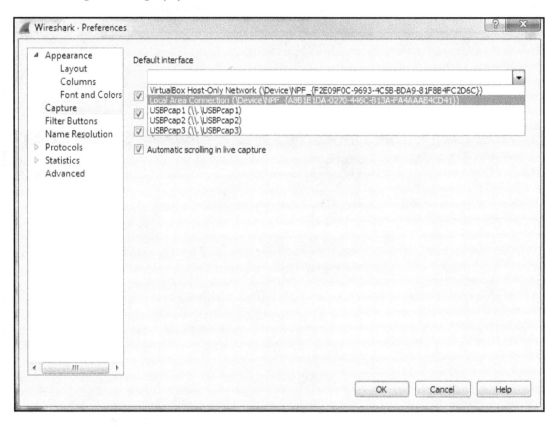

Now, as you've noticed before, I have a number of interfaces on my system. You can select your favorite interface, or your most commonly used interface on your capturing system, if you use one so very often.

You can also turn on and off the `pcapng` format, but I would highly recommend that you leave it enabled since that is now the new standard format; but if you have a requirement to capture only in the old `pcap` format for some reason, for some legacy software or something like that, you can certainly do so.

To improve performance in Wireshark, you may want to turn off these two options: **Update list of packets in real time** and **Automatic scrolling in live capture**. You've noticed what we've done so far in the captures. The moment you click on **Capture**, it starts scrolling through in that packet list, showing you everything that's coming in at that moment. Now, that is useful for small captures and quick ones, but if you have a system that's receiving a lot of data (maybe it's a SPAN port on a heavily used trunk, or the system is old and it potentially could drop packets because it doesn't have the processing power to do that) you may want to turn this off to preserve your performance in Wireshark.

Filter buttons

Your next section in **Preferences** is **Filter Buttons**:

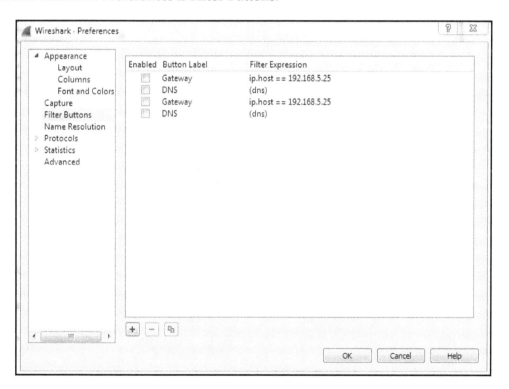

This is where all of these expression buttons will show up.

Name resolution

Our next category is **Name Resolution**, and Wireshark allows you to resolve many of the different addresses that we see in Wireshark into different names, to make it easier for us as humans to understand what we're looking at:

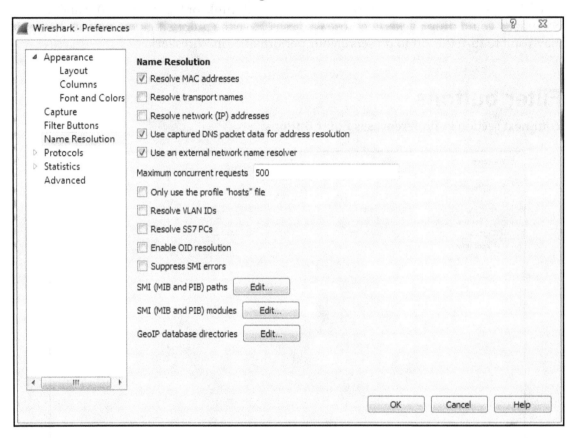

So, by default, it allows for resolution of MAC addresses to the first half of the MAC address. If you know about MAC addresses and how they work, the first half of the MAC is the manufacturer of the network card. So, Wireshark has a built-in list of these known manufacturers and the OUIs, which is the first half of the MAC, and it will try to resolve them for you. And that's why you'll potentially see `Realtek:` and then the other half of a MAC address or `Cisco:` and the last half of a MAC address. That's because of this checkbox.

You can also resolve the transport names, which are the TCP and UDP ports and IP addresses. Now, if you choose the **Resolve network (IP) addresses** option, note that it does not reference a static file within Wireshark, such as the MAC address and the transport names. It will attempt to do a DNS resolution while you're capturing. This can be a very negative thing with Wireshark, especially if you are doing a large capture with a lot of data coming in. You could have potentially thousands and thousands of DNS resolution requests going out from your capture system, clogging up the work. What I would recommend is, when you have a capture, you can right-click on an IP address and resolve it with that specific IP address rather than resolving everything. Now, the lower section, where it says **Enable OID resolution** and **Suppress SMI errors**, is for SNMP resolution. In SNMP, you have MIBs, which are basically word translations to OID locations, and you can resolve these in Wireshark.

If you are capturing SNMP traffic, you can resolve these OID strings into the MIBs if you enable the OID resolution.

Protocols

Our next category is **Protocols**, and when you expand the **Protocols** category, you have a huge list of all the protocols supported by Wireshark, and all of their associated configuration options that you can tweak.

Now, most of these you can leave alone at their defaults and everything will work just fine. There are two that you're probably going to want to tweak at some point in your career, and that'll be IP and **TCP**; or three if you count **IPv6** now. **IPv4**, **IPv6**, and **TCP** are probably the most common ones that you're going to adjust, if you adjust them at all. What we'll do is, we'll go to **IPv4**, and you'll see there's a checkbox called **Validate the IPv4 checksum if possible** that's actually disabled by default:

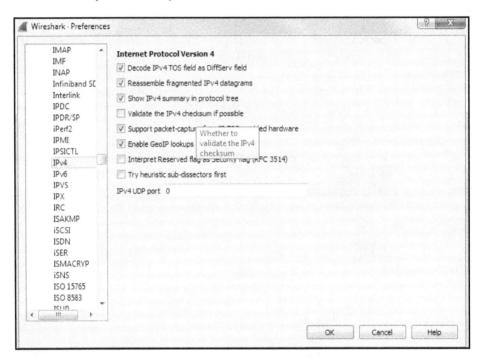

Now this used to be enabled by default, so depending on the version of Wireshark you're running, if you are not using the latest version of 2.0, the **Validate the IPv4 checksum if possible** would potentially be enabled. When that is done, it would sometimes show up based on your system with a whole lot of bad checksum errors. The reason for this is that a lot of newer systems, especially servers, have been starting to do checksum offloading where the software does not do the checksum creation but the hardware does, right before it gets sent onto the wire. But Wireshark didn't see that, so it always thought that the checksum didn't match because it couldn't see the hardware creating the checksum as it got put onto the wire. This is one thing to go into check as you most likely will want to have **Validate the IPv4 checksum if possible** off nowadays due to most network cards doing checksum offloading.

Statistics

Our next category is **Statistics**, and there's not much in here that you'll want to change:

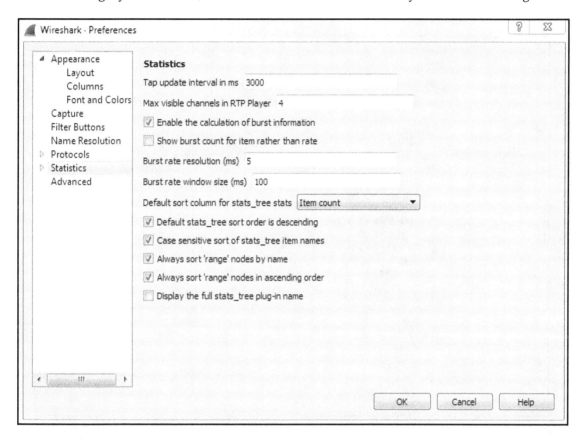

I would leave most of this as it is, unless possibly you'd want to change the number of channels in the RTP Player.

Advanced

Our last category is **Advanced**, where we have a listing of all of the preferences and settings within Wireshark, in a nice big list for you:

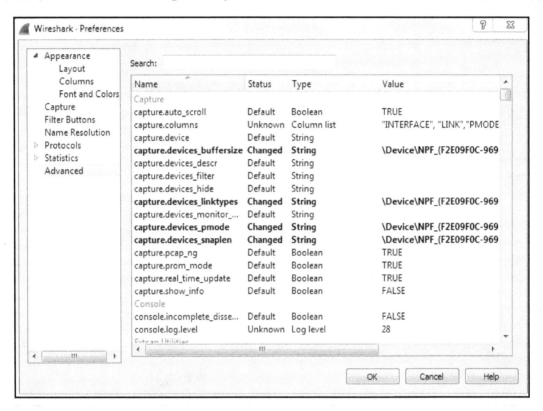

So if there's something that you needed to change but couldn't, or maybe you had a problem or something like that, and you found an answer online to change a value and you don't know where that is within the interface, you could make such changes by going to the **Advanced** category. And what's nice is there is a **Search** function. If you need to change something, you can filter it and determine where a certain setting is.

Profiles

We'll now take a look at how to create profiles to package these preferences into usable profiles that you can switch between, based on the situation that you are in.

When you're using Wireshark, any of the changes that you make to it, whether it's your preferences that you might be changing, display filters that you might be creating or capture filters, or any of that, they all go under what is known as the **default profile**. And when you create new profiles, they will work as a copy of the **default profile**. Thus, it's recommended that you make minimal changes to your default profile. You can maybe make a few overreaching changes to your environment, but don't do anything specific, and instead make a profile for different specific situations that you might need. You can do that in the bottom right-hand corner of the Wireshark interface. As you can see, there's the **Profile: Default** selected there, and if you click on that it'll allow you to select between the different profiles that you have on your system:

By default, you have a **Classic** and a **Bluetooth** profile that's included in Wireshark. You can see we're currently using the **Default** profile. If you wish to manage these profiles and create them, you can right-click on **Profile: Default**. And you'll now see a new window that pops up and allows you to manage your profiles or create one:

They take you to the same spot, though. So what we'll do is we'll just go into **Manage Profiles...**, and you can see the listing of profiles that we currently have. To create a profile, we just need to click on the plus sign:

Alternatively, if you were in that previous window, you could simply click on **New...** and it brings you to the exact same window; but, instead, it automatically clicked on the plus sign for you. What we can do is name our profile here. We'll call this New Profile, and you'll see that it says **Created from default settings** on the right-hand corner:

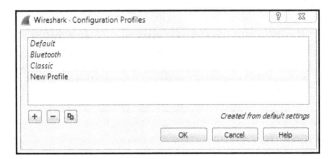

We can see how it copies the default settings, whatever you had already configured, in your system. Now, it's creating a new profile for us, and if we click on **OK**, it will create it; you see something's changed in the interface, and in the bottom right it says **Profile: New Profile** and we're now using that:

What you can do is right-click on **Profile: New Profile** and go to **Edit...**, and you'll see it has the path to the profile:

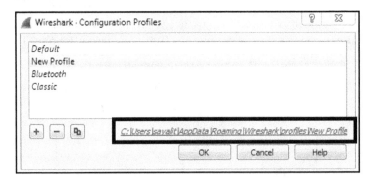

Wireshark stores profiles as folders, so click on that link and it will open up the New Profile folder. Then, we'll go back to the profiles folder and you'll see that under the profiles folder, every new profile you create will show up as a new folder:

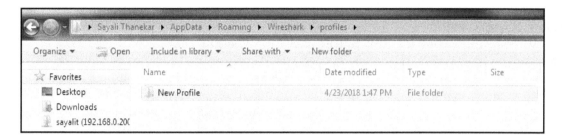

And if you go back even one step, that is in the Wireshark folder, you'll see the default files that have been created in your Wireshark installation, and there'll be different ones depending on what you might have changed:

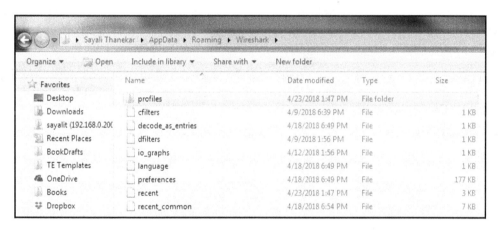

For example, there's cfilters, which is your capture filters, and we can show that. Remember that these capture filters are under your **Capture** options, and if you click on this bookmark, it's an easy way to get there. You can then manage your capture filters. Remember, I made a custom one at one point previously. That's why it has that saved. Additionally, we have io_graphs preferences; we have the language selections now that Wireshark is in multiple languages; and we have the recent files and the preferences, as we were just using in the previous section. And another one would show up if I'd created display filters; there would be a dfilters file too, which would have customized display filters. Now what we can do is we can edit these files as well. These are text files. What you can do is right-click on any file and edit it.

I would recommend using something such as Notepad++ that will show the carriage returns correctly because if you open it with Notepad, it might not show up correctly due to the type of carriage returns they use.

Now, if you wish to share your profiles with other people, you can go into the **profiles** folder and simply copy the **New Profile** folder for whatever profile you wish to share. Maybe you have an **802.11** wireless one, or a TCP analysis one, a corporate network one, or a major errors one-whatever it is that you might have-and you have multiple IT administrators or analysis individuals that are in your organization. You can share these profiles among each other by simply copying and pasting these folders between your different computers, and you could share it as a ZIP file or whatever suits you best.

Colorizing traffic

In this section, we'll take a look at how to create coloring rules, how to remove them, how to make your own, how to colorize conversations, and filter traffic by the coloring rules that are applied. Let's start.

I pulled up one of the previous captures that I've done—loading up a web page. We'll use this as an example for our coloring rules discussion. You can see here that it's already been colorized by default from Wireshark:

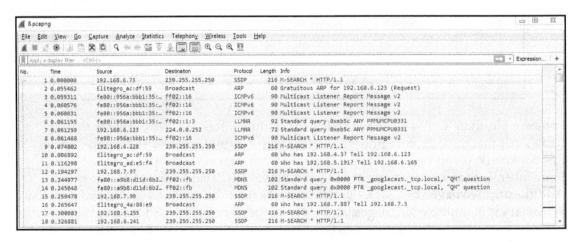

As I mentioned before in the profiles in the **Preferences** section, there are **default settings** within Wireshark, which include the coloring rules. Wireshark tries to make it a little bit easier for you to view your traffic, even from just a standard install. But you can, of course, customize that to show what you want to see.

The first thing you want to know, besides the fact that Wireshark will automatically colorize your traffic and from a basic standpoint, is that you can enable and disable the rules with one button. Up at the top of the window, you can click on the **Draw packets using your coloring rules** button that looks like kind of a rainbow color of lines, and you can click on this to enable or disable the coloring rules in bulk:

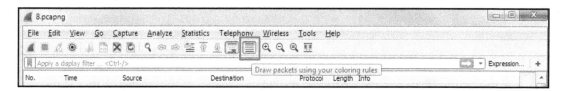

If I click on this and disable, now everything turns to white and there are no coloring rules currently applied:

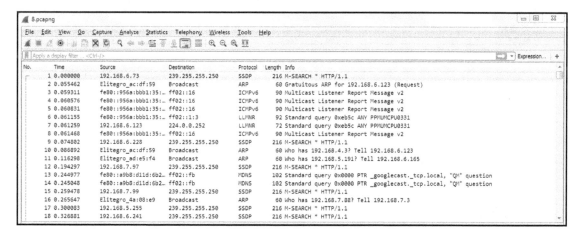

If you click on the **Draw packets using your coloring rules** button again, it will re-enable the coloring rules.

In order to get to the coloring rules, we'll go to **View** | **Coloring Rules...**:

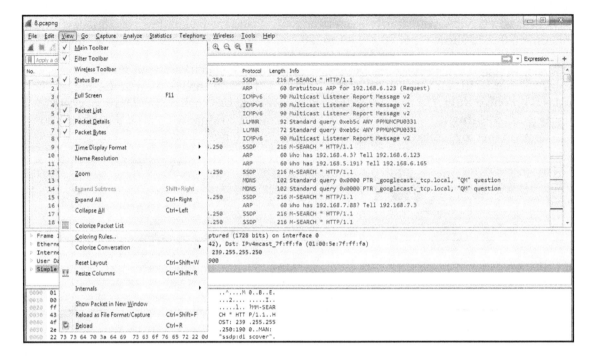

Clicking on the **Coloring Rules...** option will bring up the **Coloring Rules** window, and you can see that the default selection of coloring rules are created by Wireshark at installation:

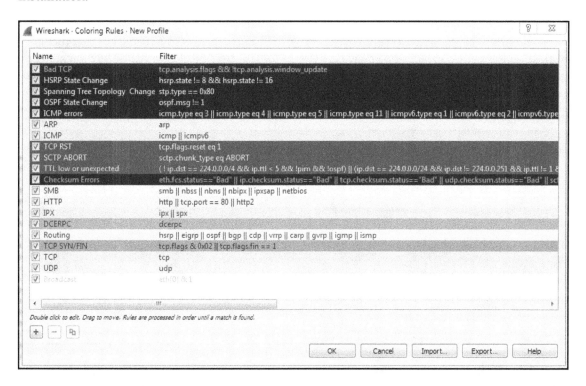

You can see how the default rules have some real vibrant coloring rules right at the very top. And these are some of the nasty ones that you want to keep an eye out for, and that's why they have a black background. That way, they stick out in this default pastel coloring system that they have for the other traffic. You can see most of the other normal stuff, such as **ARP**, **ICMP**, **SMB**, **HTTP**, and **UDP**; all of them are nice pastel-type colors. And then all of a sudden you have this black background show up with red text and you know that's obviously something bad. You have the yellow show up; you have the bright red with the TCP resets and the aborts; and you have **Checksum Errors** that are black with a red text. They're supposed to stand out on purpose. Another thing to know about this is that you see how the bad ones are up at the top: the rules are processed from top down, so it's based off of a matching system. If something matches first, then that will be applied. If you have two rules that are very similar and, maybe you have IC-two rules that look for **ICMP** but have different colors for some reason-the one at the top will be applied first and the one after it will be skipped as it has already been colorized.

Additionally, on the left-hand side of all these coloring rules are a bunch of checkboxes, and these checkboxes enable and disable these individual coloring rules. That way, you don't have to enable or disable all of them at once with that button I just showed you. Thus, you can simply turn off the checkboxes you want or don't want.

For example, what we can do is, we can turn off **HTTP**, which is this nice light green color you can see throughout all of our capture if you look at the scroll bar. All of our packets are green. So I'll turn **HTTP** off; we'll then click on **OK**. When I do this, you can see that they all turn into purple:

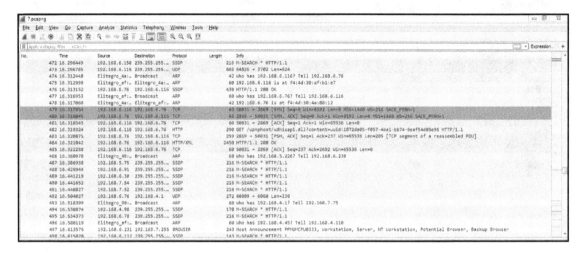

Now why do they turn into purple? Let's go back to **View** | **Coloring Rules...** and take a look. And if we check, the TCPs are light purple. So here, all the communication, all the packets, are TCP packets.

Remember I was talking about how things match first on this list towards the top? So, obviously, the HTTP traffic, which is more specific since that's a higher-level protocol, matched first and so was applied green. And then, when it matched for TCP it was ignored because it was already colorized. You can see how you want to put more specific things towards the top and then more generic things towards the bottom, and you can see how it's kind of broken out that way already with all these very specific filters up at the top and then the basic ones at the bottom, such as **TCP** and **UDP**.

You can also create your own, of course. In the bottom left, just like the other windows we've seen already, you can click on the plus sign that will create a new coloring rule. And you can see it puts it up towards the top for you, as well, which is very nice. Here, you can enter any sort of display filter you want to create a color. We could say `ip.host` is my gateway and we can choose a foreground and background color. You can see at the bottom that we can choose a **Foreground** and **Background** color:

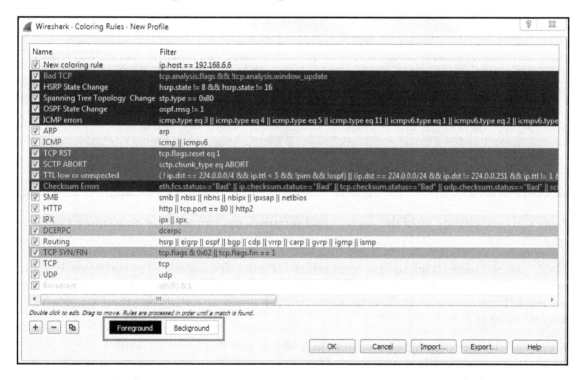

So, let's make it something completely ridiculous. The foreground will be bright pink and the background will be a hideous yellow color. Let's rename it first. We'll say `Gateway`, and click on **OK**. That definitely stands out, doesn't it? Check out the following screenshot:

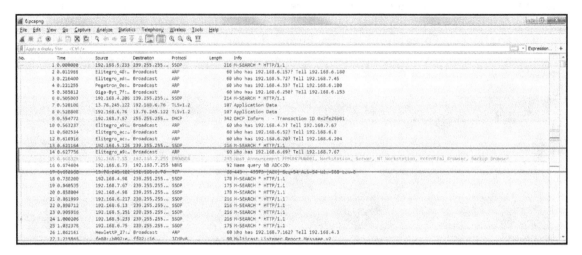

Now you can see how all of the packets that reference my gateway are now colored in bright yellow.

Until now, we were trying to figure out which coloring rule applies to a specific packet, and we were doing so by going to **View | Coloring Rules...** and trying to just figure it out by ourselves by going down the list. Now, you don't have to do it that way; there's an easier way of doing it.

If you click on a packet that you want to investigate and find out why is this green, we can look in the packet details and, under the **Frame** section we can go down towards the bottom of the frame information, and it will show us the coloring rules that are applied. So, the coloring rule name is **HTTP**, and the string that applies to our selected packet is **http ||** **tcp.port ==80 || http2**:

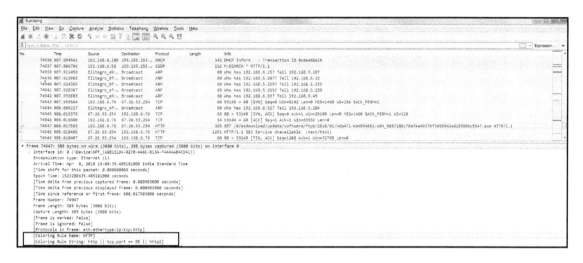

Now, if you make a whole bunch of custom coloring rules for your specific company in your certain situation that you have going on, you can absolutely import and export these to share them with co-workers. Let's go back to **View** | **Coloring Rules...**, and you can see in the bottom right of the window that we have **Import...** and **Export...** buttons, which will export it or import it as a text file:

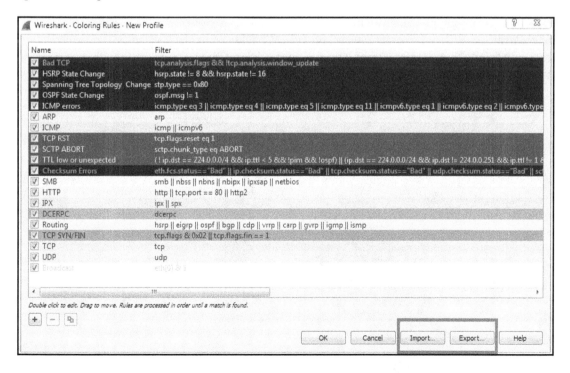

Additionally, we can colorize conversations. What we'll do is, we'll go and select a random packet in this conversation, and we can right-click on it and go to **Colorize Conversation**. And then we can break down on which conversation we want and which layer of the OSI model:

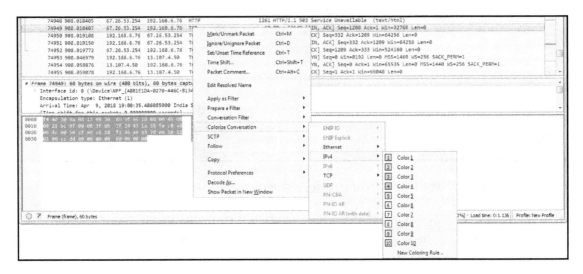

And finally, one cool feature I want to show you is you can go ahead and actually filter on the coloring rules. Let's select a packet, an **HTTP** packet, and we'll go into that frame section in the packet details again. And I'll right-click on the coloring rule that is applied to this packet and I can go to **Apply as Filter | Selected**:

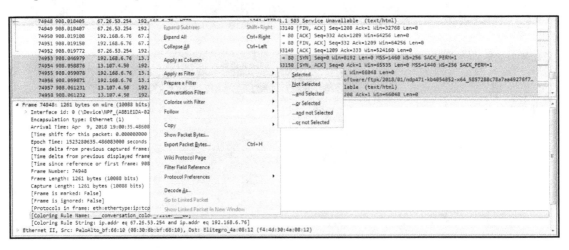

Now, I'm filtering on that coloring rule, so it's not just on the fact that it's HTTP traffic, but it's on the coloring rule applied.

So we can do that to another one as well. We'll go and right-click on the coloring rule that is applied to the selected packet, and we'll select **Apply as Filter | Selected**. And now we've just filtered out that conversation based on the coloring rule.

Examples of colorizing traffic

We will go through some examples as to how you can use coloring rules to your advantage to pick out bad things in different packet captures.

Example 1

In this example, we will need to open up a web page that is to a nonexistent location on MIT's website, so there should be a response code from HTTP in here stating 404, that it couldn't find that file. What we'll do is create a coloring rule that will easily show that the error message in the packet list will pop out to us:

1. Let's go to **View | Coloring Rules...** and we'll create a custom coloring rule for HTTP response codes by clicking on the plus icon.

2. Type in `http.response.code > 399` in the **Filter** column:

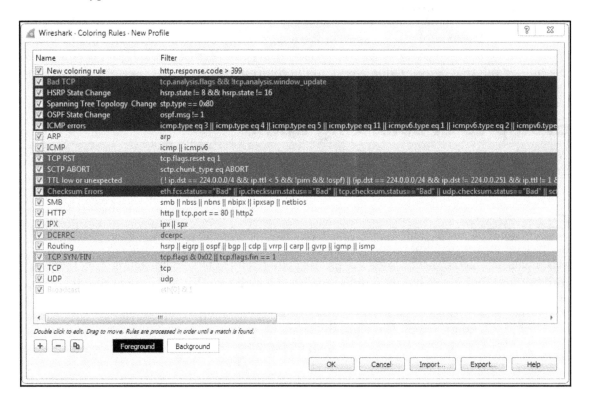

 All of the HTTP response codes that are above 400 are server and client errors, in the 400 and 500 range respectively.

3. Since we want to see all of the errors in a very vibrant color, we will make the **Background** bright pink and the **Foreground** black. You'll see that's right up at the top there in our list, so that'll be applied first. Let's click on **OK**:

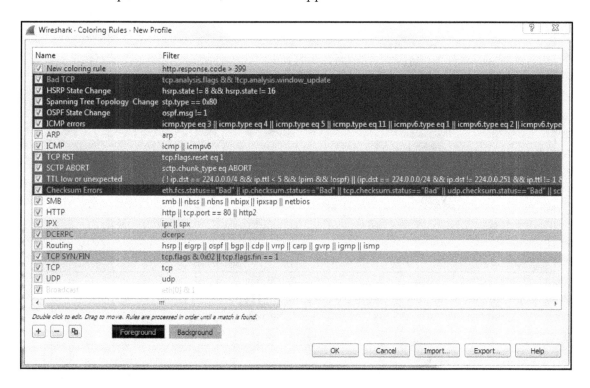

4. We'll scroll down through our list looking for that bright pink packet. It pops in real easy and is very visible to us. So, as you can see, it says **HTTP/1.1 408 Request Time-out (text/html)**:

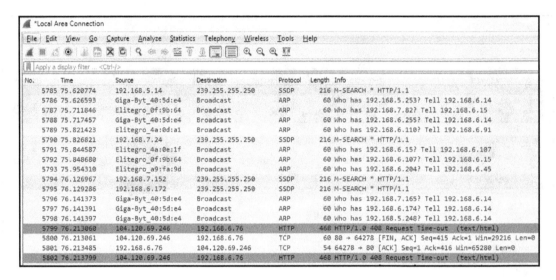

5. We can expand it in detail by clicking on **Hypertext Transfer Protocol** at the bottom and take a look at that. We'll then see the status code of `408`:

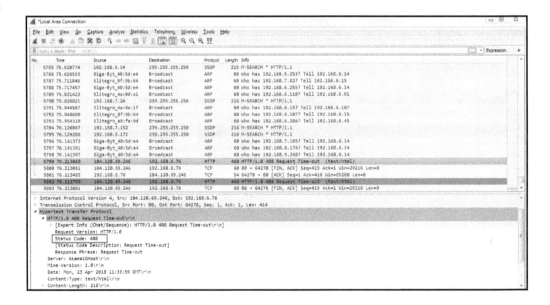

Example 2

Now, in the second example, we'll take a look at NS. Let's say, you have some sort of problem in your network where certain **DNS** responses are coming back if they can't find any sort of records in your **DNS** server for a resource, whether it's a web page or a local server; whatever it is, it's not finding the **DNS** entries. We can apply a coloring rule that will vibrantly show us whenever there's a **DNS** response that it cannot find a record.

Let's go ahead and create a new capture, and I'm going to look up some random web page that doesn't exist, and we'll use **DNS** off of Google. We will get a message that it doesn't exist:

```
C:\Windows\system32\cmd.exe

Microsoft Windows [Version 6.1.7601]
Copyright (c) 2009 Microsoft Corporation.  All rights reserved.

C:\Users\sayalit>nslookup 123456wertwedvg5.com 8.8.8.8
Server:  google-public-dns-a.google.com
Address:  8.8.8.8

*** google-public-dns-a.google.com can't find 123456wertwedvg5.com: Non-existent
 domain

C:\Users\sayalit>
```

So we're going to stop our capture, and we'll see nothing really pops out. There would be some red stuff there from some TCP resets and some black ones, but nothing else. It's all TCP traffic and **DNS**. Let's go ahead and create a coloring rule:

1. Make a new coloring rule by clicking on **View** | **Coloring Rules...**; we'll call it
   DNS No Records.

2. In the filter information, we'll enter `dns.flags.rcode`, which is for the response code. If the response code equals 3, it means there's no record:

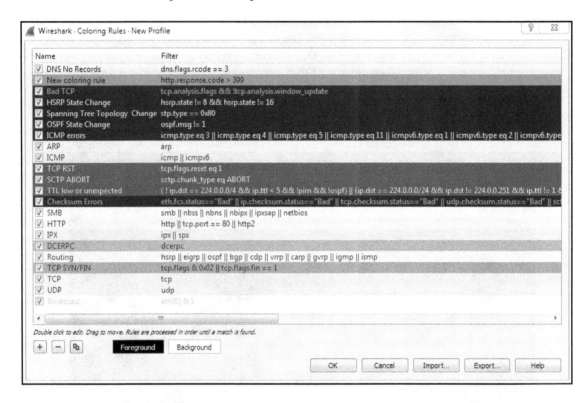

3. We will select the **Foreground** as black and **Background** as yellow.

Now, let's go ahead and scroll through our packet capture and see if we can find this response code error. We can see our `Standard querie response 0x0002 No such name A` record, and then the random gibberish that we entered in. And then of course it's also responding for IPv6. The `AAAA`, the quad A record, is for IPv6, as well:

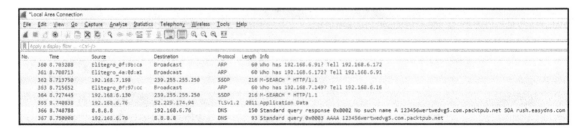

If you take a look in our frame information at the bottom, there's our coloring rule: DNS No Records. There's a string that it's using, and if we dig into the DNS information, you look at the flags and it will say Reply code: No such name (3). That's the value of 3 that we just filtered on:

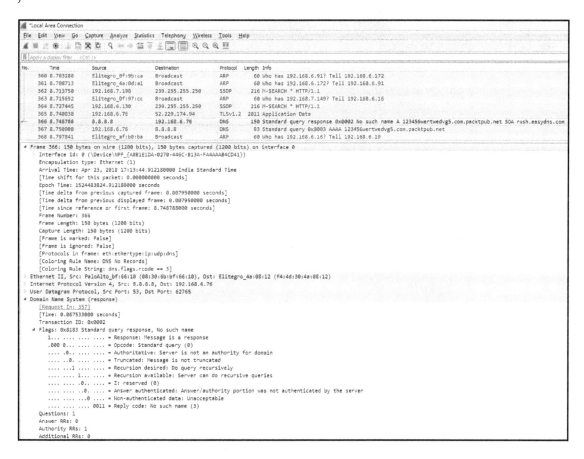

You can see how applying coloring rules can be very helpful to someone who's doing network analysis and protocol analysis because it makes it very easy to pick out from a large packet capture wherever there's a problem.

Summary

In this chapter, we've learned how to apply preferences in Wireshark and customize it to your needs. We've also learned how to create profiles for different analysis requirements, and switch between them. For example, the coloring rules that we just saw could be put into a profile specifically for DNS or HTTP. And we've learned how to create these coloring rules, how to import and export them, and how to apply them to real-world examples.

In Chapter 5, *Statistics*, we'll dive into statistics within Wireshark, which is a great feature that very few people seem to use.

5
Statistics

In this chapter, we will cover the following topics:

- TCP/IP overview
- Time values and summaries
- Trace file statistics
- Expert system usage

TCP/IP overview

In this section, we'll take a look at the basics of TCP/IP, how packets are built, and the resolution processes that are in place, such as DNS and ARP.

In networking, we have two models that we commonly use: OSI and TCP/IP. As shown in the following diagram, on the left side we have the OSI model and on the right side we have TCP/IP model, and I've tried to match them up so that you can see how the different layers of each model line up with each other:

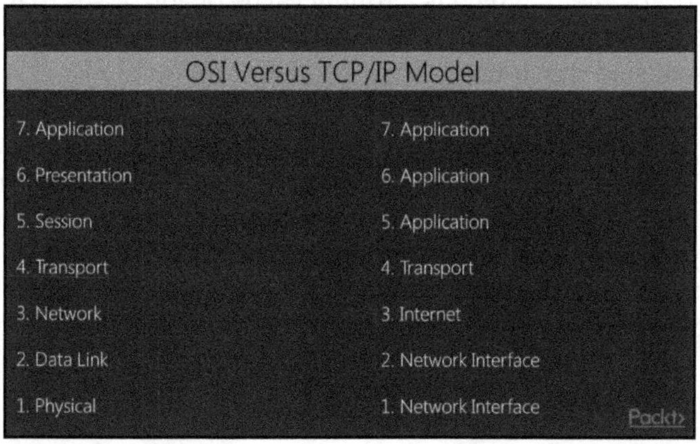

OSI Versus TCP/IP Model	
7. Application	7. Application
6. Presentation	6. Application
5. Session	5. Application
4. Transport	4. Transport
3. Network	3. Internet
2. Data Link	2. Network Interface
1. Physical	1. Network Interface

When we use Wireshark, we're commonly concerned with layers 2 through 7 of the OSI model. And most commonly when you use Wireshark, it's probably because something that's often application-related is going on or the system is running an application. Most commonly, you'll find yourself using Wireshark to diagnose problems that are in the upper layers, especially layer 7. But you can certainly use it to troubleshoot connectivity issues between devices on layer 3 or layer 2. While there are a number of TCP/IP services and protocols that we use to help us communicate over a network, note that we reference the layer where that protocol resides based on the OSI model, not the actual TCP/IP model's layer.

What I'd like to do is run through the building of a packet, which will give you an idea as to how the values are entered into a packet for the different fields. And since we're looking at these fields in Wireshark, it's certainly a good thing to know. So what I've done is, in a browser, I've opened up a connection to `http://www.pbs.org/`. What we'll do is follow through that connection and show how it found the resource and then send the first data packets to it.

The first thing that your system needs to figure out is what port number to use. When you open up your web browser and go to `http://www.pbs.org/`, depending on how you enter the address into the address bar of the browser, the application of the browser will know whether or not you want to use port `80`; it would know by default for HTTP, or port `443` for SSL, or maybe some other custom port. So right away, your computer knows what port number it needs to start communicating on. Since I went to `http://www.pbs.org/` without any sort of SSL connection, it by default knew that it will have to use port `80`. The next thing it has to do is figure out where that service is. So I went to `http://www.pbs.org/`, but my system didn't know where `http://www.pbs.org/` resides.

DNS deals with the resolution of a name to an address. So my system had a look at the DNS cache on my local system, which is a rolling cache of addresses that it has already resolved, and it looked for `http://www.pbs.org/`. It saw that `http://www.pbs.org/` didn't exist in the cache, so it said that it needs to go send that out to my DNS server to hopefully get a response as to where this resource is located. My DNS server happens to be on my gateway, which is `0.6`; this is common for standard home networks. It may not be there; it could have been a remote resource as well, such as a Google DNS server or open DNS, or some other. If this is a remote DNS server that I'm trying to connect to, then my system will take a look at my route table and figure out where it needs to go in order to access that DNS server. So, if it's remote, outside of my network, it's going to take a look at my route table and realize that it has to go out through my gateway in order to go talk to the DNS server.

And when it does, it'll check my ARP cache to see if there's a layer 2 address for my gateway so that it can send a frame to the gateway. If my system didn't have an ARP cache entry for my gateway, then my system would have sent an ARP packet out, looking for the physical address for `192.168.0.6`. My system happened to have it in its ARP cache since I commonly accessed that IP address, and so we don't see an ARP packet here. What we can see here is that the first packet is a DNS packet, so my system saw that `0.6` was my DNS server, and it knew that it was a local resource. It checked my route table and realized that it's on my physical interface there and we're already connected to the `0` network, and also that I already have an ARP cache entry for `0.6`. So it didn't need any of that information, and it automatically built the DNS packet. We can see in here in IPv4 that my destination is `0.6`:

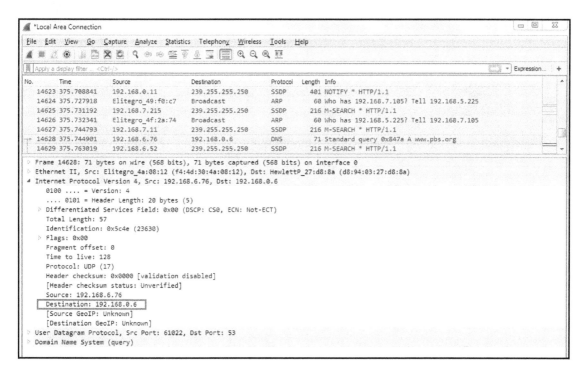

It built all that without having to produce any other packets to find this information. So my system then sent out its DNS request asking for `http://www.pbs.org/`. Now, if we go down and take a look at the UDP section, you can see that we're using port 53, which is for DNS:

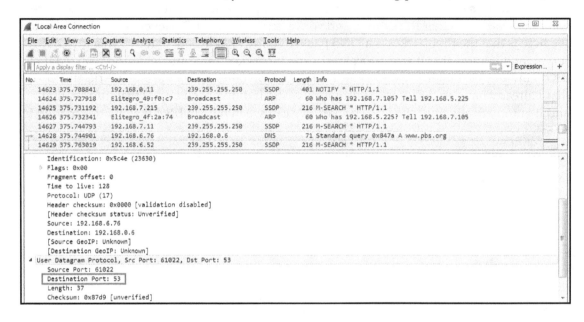

And if we expand DNS, we can look at the **Queries** and see that we're asking for `http://` `www.pbs.org/`:

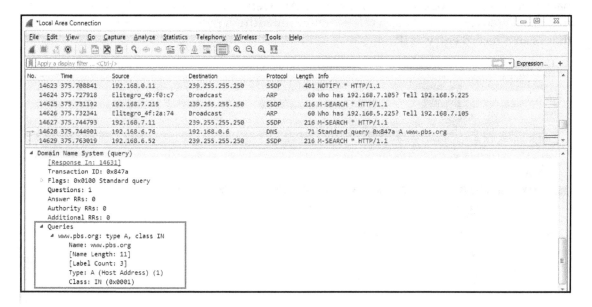

We then have to wait for the DNS response. Then, we see the next packet for DNS. We receive a standard query response for `http://www.pbs.org/`. If we look at DNS in the packet details pane, we can take a look at the **Answers** in this packet. And you'll see that `http://www.pbs.org/` has a `CNAME` entry for address `r53-vip.pbs.org`, and then there's an `A` record entry at IP `54.225.198.196`:

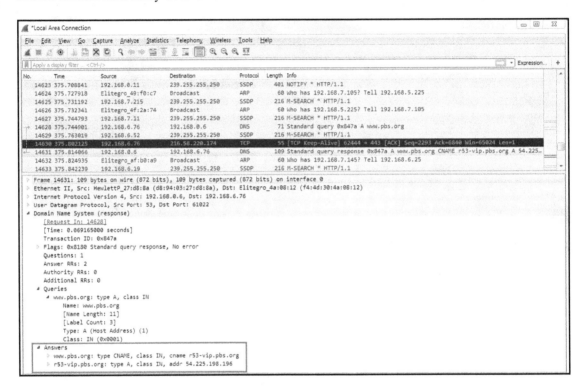

The A record entry is what we care about. So now my system has an IP address for `http://www.pbs.org/` so that I can build a TCP handshake to `http://www.pbs.org/` and to the web server so that it can then begin flowing HTTP data back and forth. So if we look at the next packet in line, we have our SYNs:

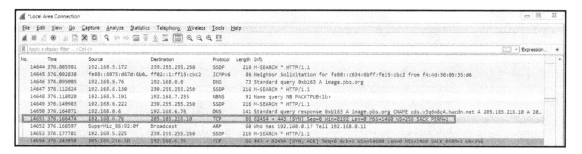

So, the highlighted packet in the preceding screenshot is the first packet in our three-way handshake for TCP. If you're familiar with TCP at all, it goes SYN, SYN, ACK, and ACK; that's your three-way handshake to open up a connection. So my first packet has a destination address of 205.185.216.10. If we look at that response that came back in DNS, the address that was provided for the A record is 205.185.216.10. So now my system knows what IP address to craft its TCP packet for. We now have a source of my local system and a destination to the server for `http://www.pbs.org/` that it received, and then it sends out its handshake request. It then gets a SYN, ACK in response saying yes, I see your connection requests; let's create a connection. Then my system responds, saying yeah, that sounds great; I acknowledge. And then, finally, we begin our first HTTP packet. My system sends an HTTP packet out to that same server and says GET / HTTP, so it's saying please send me your beginning index.html and any other data for the HTTP resource. And again, it knew that it needed to use port 2869 HTTP unencrypted because when we entered that into the browser, the browser application let the stack know that it was using port 2869. And so that's the basics on how a TCP packet is built, and your system requests a resource from some sort of server or other device out there.

Time values and summaries

In this section, we'll take a look at how to change the time settings for packets and troubleshooting with the time column.

We have the PBS packet capture again, where I opened the browser and went to `http://www.pbs.org/`. If you notice in the packet capture, the second column says **Time**:

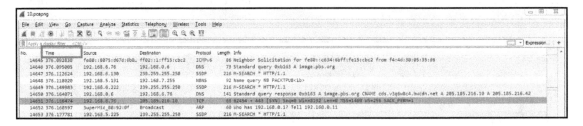

The **Time** column is a number with a decimal and it just keeps counting up as you scroll down through the packet capture. By default, in Wireshark, this is the time since the capture started. Having the time since it was captured can be useful so you know when certain packets arrive in relation to the entire data flow that you captured, but it's not all that useful for trying to diagnose a problem where there might be a delay in a certain service returning traffic that you're trying to capture back to your system.

In order to figure out the delay between captured packets, you'd have to look at the **Time** column and figure it out based on milliseconds, microseconds, and nanoseconds, and that's not that great for humans because we're not all that great at math to that level. So what we can do instead is go to **View | Time Display Format**, and we have a large selection of time display formats we can choose from. And the most useful one that I would recommend is using **Seconds Since Previous Displayed Packet**:

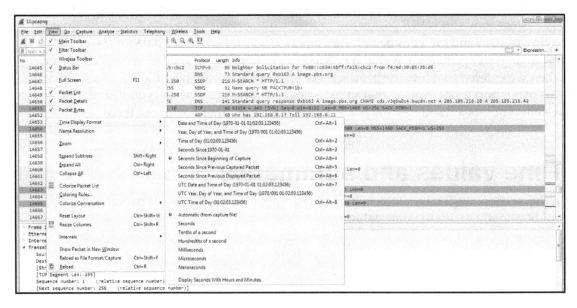

This way, if you apply a filter on your traffic, such as following a TCP stream, it will show you the delta difference between each packet based on the applied filter. If you use **Seconds Since Previous Captured Packet**, then if you have packets that are filtered out from the view that you're looking at, it'll not coincide exactly with what you're looking at; it makes it a little bit harder to understand. So what I would recommend is choosing **Seconds Since Previous Displayed Packet**.

Now if we look, the **Time** column has changed and we have 0 seconds between each packet, and then there's some sort of fraction of a second for each packet for the delay between each captured packet:

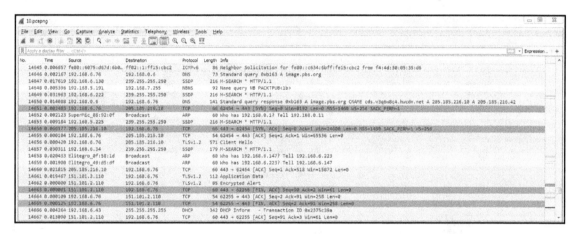

What we'll do is we'll scroll up to the top and you can sort the **Time** column. And if you sort it by highest to lowest, you can see what packets are the most delayed. Now, if you select that packet and then re-sort by the **No.** column, you'll be taken directly to that packet, wherever it is within the numbered packets that you've captured. You can look on either side of the packet to figure out what might be going on.

You can also add additional columns. So, what we can do is go to packet details and expand the frame information. And you'll see there are a number of time fields. What we can do is also add a column for one of these time fields. So, what we'll do first is we'll switch our display back to since the beginning of the capture by going to **View | Time Display Format format | Seconds Since Begining of Capture**, and we'll add a column for the delta between each displayed frame. We'll do that by selecting **Time delta from previous captured frame** from the frame information. We'll right-click and select **Apply as Column**:

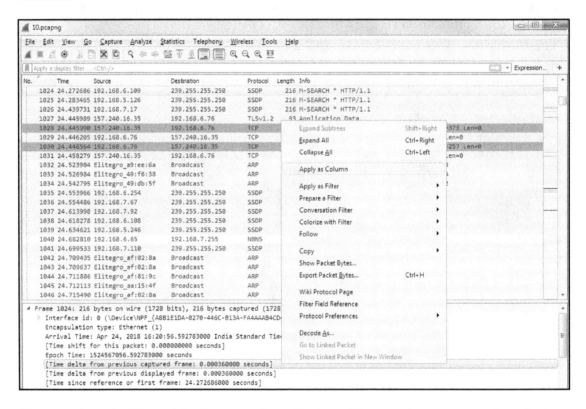

Then, we drag the **Time delta from previous displayed frame** column over to the other time so that it makes it a little bit easier.

So, you have time since the capture began, and then time delta between each displayed frame. You can also go to **View** | **Time Display Format** and change the fraction of seconds based on what you want to see. So, maybe you don't need to see the nanoseconds, and you only care about the milliseconds. You can change that manually by selecting the **Milliseconds** option:

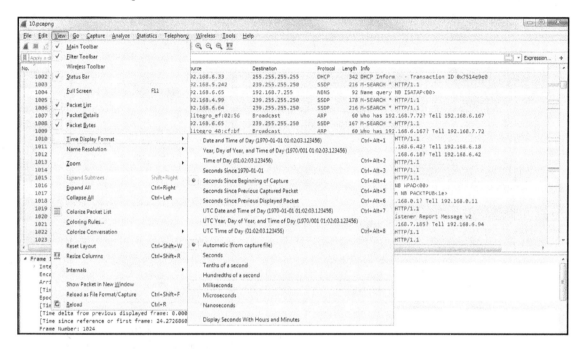

You will now be able to see how it pruned our **Time** column.

You can also add columns by going to **Edit** | **Preferences...** and then to the **Appearance** | **Columns** area, and you can manually add whatever column you want by clicking on the plus sign:

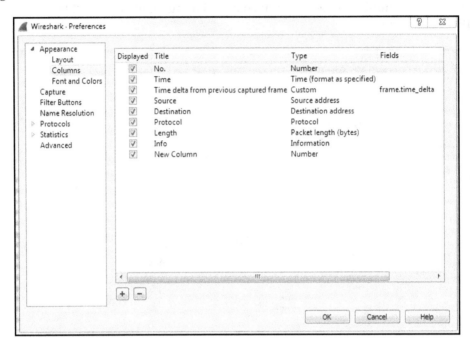

Lastly, if you go to **Statistics** | **Capture File Properties**, you'll see a list of information based on the packet capture. And if you scroll down, you'll notice that there's a whole bunch of statistics on the capture itself, including the number of packets, the time span, how big this packet capture is in number of seconds, average packets per second, average byte size, total number of bytes that have been captured, average bytes per second, and average bits per second:

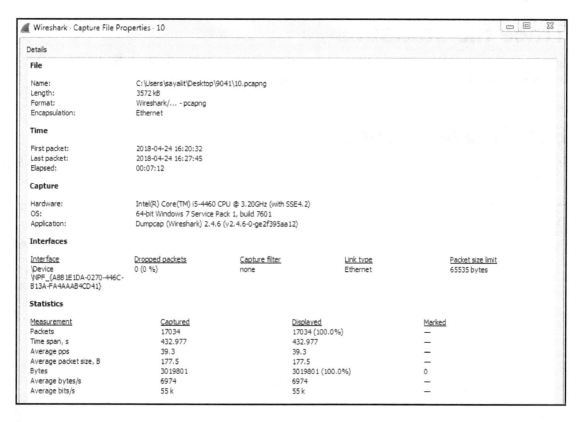

This summary can be useful when comparing a capture done during a benchmark, when everything is running normally, and compared to a packet capture performed when there are performance issues. You can see if there are any of these values in the summary statistics that have changed drastically.

Trace file statistics

In this section, we'll take a look at how to display useful statistics in Wireshark and some issues you could troubleshoot utilizing that statistical information.

Resolved addresses

In order to access the statistics in Wireshark, click on **Statistics** and go to **Resolved Addresses**:

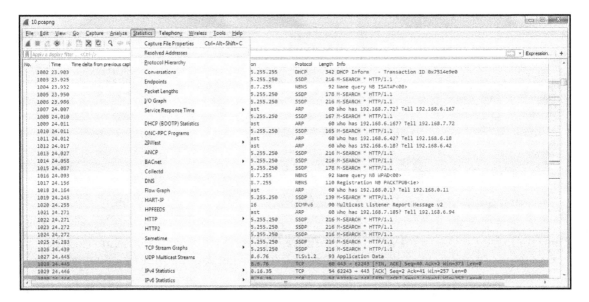

The **Resolved Addresses** window will give you a list at the top of all of the IP addresses and DNS names that were resolved in your packet capture. This way, you can get an idea of all the different resources that were accessed in your packet capture:

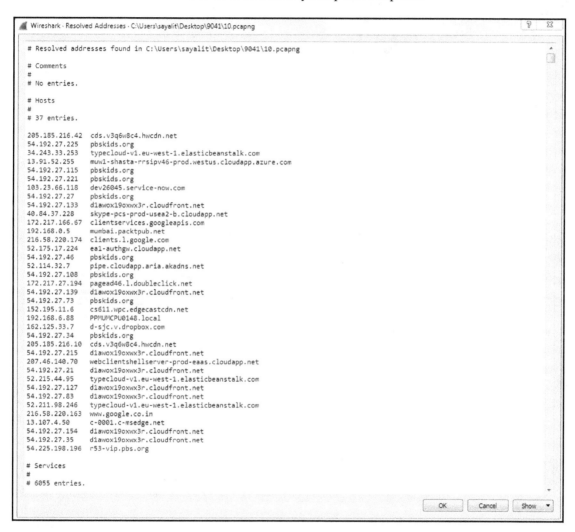

Protocol hierarchy

Next we'll look at protocol hierarchy. You need to click on **Statistics** and go to **Protocol Hierarchy**:

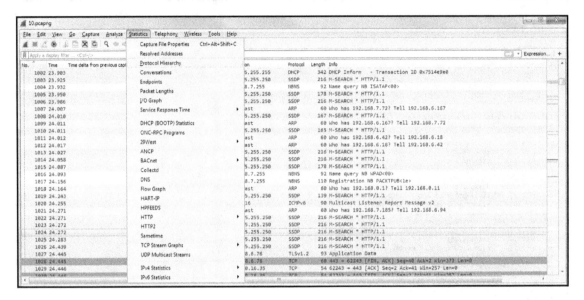

It will give you a breakdown based on the percentages of the packets of the most popular protocols that it saw:

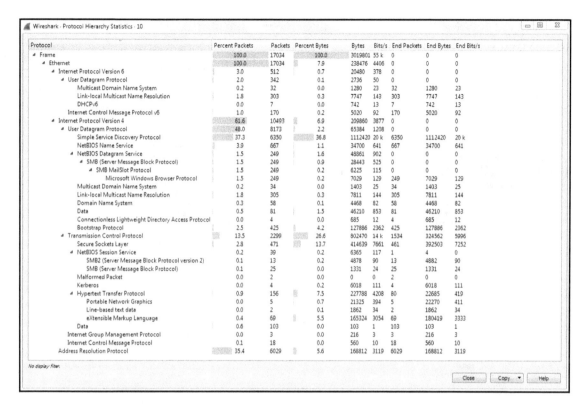

As you can see at the beginning, everything that came in was a **Frame**. Everything that came in that **Frame** was an **Ethernet** frame. And then within that, we have a breakdown of what's within **Ethernet**. So we have some **Internet Protocol Version 6**; we have a whole bunch of **Internet Protocol Version 4**, and a little bit of **Address Resolution Protocol**:

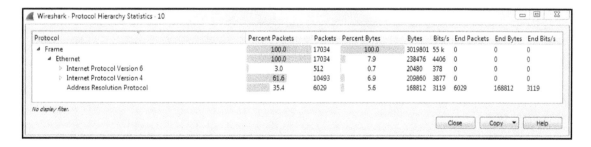

If we expand **Internet Protocol Version 4**, we see that the biggest amount of packets that we received were SSDP packets:

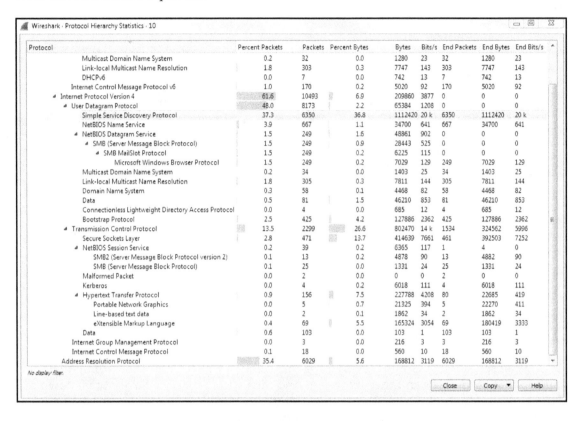

Protocol	Percent Packets	Packets	Percent Bytes	Bytes	Bits/s	End Packets	End Bytes	End Bits/s
Multicast Domain Name System	0.2	32	0.0	1280	23	32	1280	23
Link-local Multicast Name Resolution	1.8	303	0.3	7747	143	303	7747	143
DHCPv6	0.0	7	0.0	742	13	7	742	13
Internet Control Message Protocol v6	1.0	170	0.2	5020	92	170	5020	92
▲ Internet Protocol Version 4	61.6	10493	6.9	209860	3877	0	0	0
▲ User Datagram Protocol	48.0	8173	2.2	65384	1208	0	0	0
Simple Service Discovery Protocol	37.3	6350	36.8	1112420	20 k	6350	1112420	20 k
NetBIOS Name Service	3.9	667	1.1	34700	641	667	34700	641
▲ NetBIOS Datagram Service	1.5	249	1.6	48861	902	0	0	0
▲ SMB (Server Message Block Protocol)	1.5	249	0.9	28443	525	0	0	0
▲ SMB MailSlot Protocol	1.5	249	0.2	6225	115	0	0	0
Microsoft Windows Browser Protocol	1.5	249	0.2	7029	129	249	7029	129
Multicast Domain Name System	0.2	34	0.0	1403	25	34	1403	25
Link-local Multicast Name Resolution	1.8	305	0.3	7811	144	305	7811	144
Domain Name System	0.3	58	0.1	4468	82	58	4468	82
Data	0.5	81	1.5	46210	853	81	46210	853
Connectionless Lightweight Directory Access Protocol	0.0	4	0.0	685	12	4	685	12
Bootstrap Protocol	2.5	425	4.2	127886	2362	425	127886	2362
▲ Transmission Control Protocol	13.5	2299	26.6	802470	14 k	1534	324562	5996
Secure Sockets Layer	2.8	471	13.7	414639	7661	461	392503	7252
▲ NetBIOS Session Service	0.2	39	0.2	6365	117	1	4	0
SMB2 (Server Message Block Protocol version 2)	0.1	13	0.2	4878	90	13	4882	90
SMB (Server Message Block Protocol)	0.1	25	0.0	1331	24	25	1331	24
Malformed Packet	0.0	2	0.0	0	0	2	0	0
Kerberos	0.0	4	0.2	6018	111	4	6018	111
▲ Hypertext Transfer Protocol	0.9	156	7.5	227788	4208	80	22685	419
Portable Network Graphics	0.0	5	0.7	21325	394	5	22270	411
Line-based text data	0.0	2	0.1	1862	34	2	1862	34
eXtensible Markup Language	0.4	69	5.5	165324	3054	69	180419	3333
Data	0.6	103	0.0	103	1	103	103	1
Internet Group Management Protocol	0.0	3	0.0	216	3	3	216	3
Internet Control Message Protocol	0.1	18	0.0	560	10	18	560	10
Address Resolution Protocol	35.4	6029	5.6	168812	3119	6029	168812	3119

No display filter.

Close   Copy ▼   Help

Now, this is useful because you can see all the different types of data that have arrived. For example, if you're not expecting to see **Connectionless Lightweight Directory Acess Protocol** or NetBIOS UDP frames that's useful to see, especially if it were a higher percentage in the number of packets than it saw in the capture. Or let's say, maybe, you shouldn't be seeing any SMB traffic but you do see a lot of SMB traffic; that could be some sort of a breach.

Conversations

You need to click on **Statistics** and go to **Conversations**:

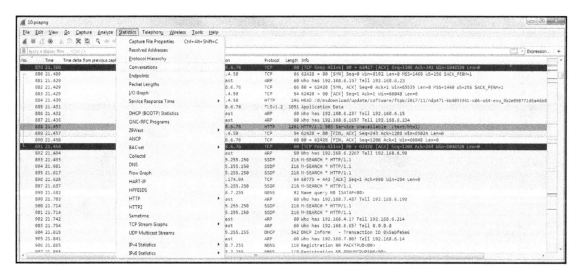

In **Conversations**, we have a list of all of the different Ethernet, IPv4, IPv6, TCP, and UDP conversations that have occurred within this packet capture:

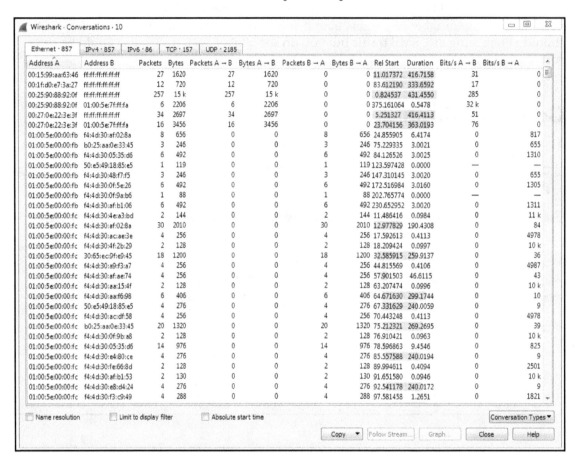

Additionally, on all of these tabs, there's a **Duration** column, and that's very useful to see which are the largest and longest talkers in your conversations which have been captured. You can sort by duration by clicking on the **Duration** column, and see which ones are sending the most amount of data:

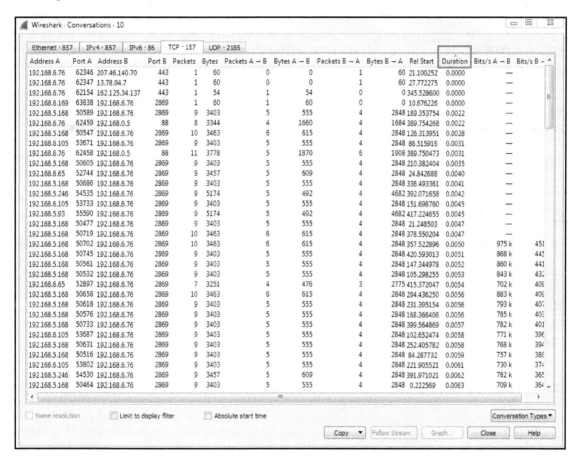

Endpoints

We'll now go to endpoints. Click on **Statistics** and go to **Endpoints**:

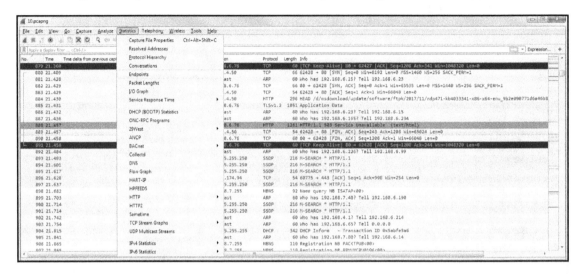

Endpoints is similar to conversations, but it's simply a listing of all of the endpoints; not just the connections between the endpoints as to which IP is talking to which IP, or which MAC is talking to which MAC; it's just a listing of all of the devices on each of the layers and the information about them. We're not concerned about the conversation, just the endpoints: all the endpoints added together.

As you can see under TCP, for example, we have `211` versus the `449` for IPv4, and that's because my host computer here opened up a whole bunch of ports when I opened that browser and tried to go to PBS:

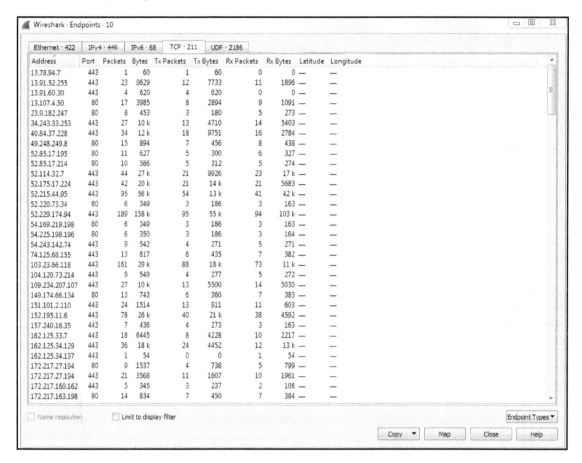

Packet lengths

Next, we'll go to packet lengths. Click on **Statistics** and go to **Packet Lengths**:

Packet lengths is useful to determine if you have small packet lengths, especially if you're having a window size issue and a lot of your data is smaller than it should be. What you want to look for is whether most of your data is in the `1280-2559` range for packet length because the maximum MTU is `1500`:

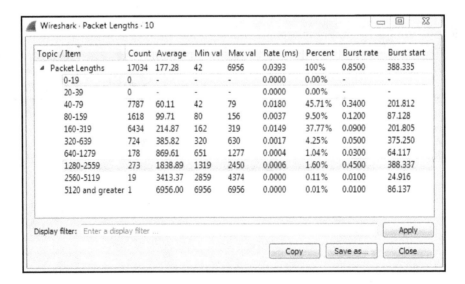

Topic / Item	Count	Average	Min val	Max val	Rate (ms)	Percent	Burst rate	Burst start
◢ Packet Lengths	17034	177.28	42	6956	0.0393	100%	0.8500	388.335
0-19	0	-	-	-	0.0000	0.00%	-	-
20-39	0	-	-	-	0.0000	0.00%	-	-
40-79	7787	60.11	42	79	0.0180	45.71%	0.3400	201.812
80-159	1618	99.71	80	156	0.0037	9.50%	0.1200	87.128
160-319	6434	214.87	162	319	0.0149	37.77%	0.0900	201.805
320-639	724	385.82	320	630	0.0017	4.25%	0.0500	375.250
640-1279	178	869.61	651	1277	0.0004	1.04%	0.0300	64.117
1280-2559	273	1838.89	1319	2450	0.0006	1.60%	0.4500	388.337
2560-5119	19	3413.37	2859	4374	0.0000	0.11%	0.0100	24.916
5120 and greater	1	6956.00	6956	6956	0.0000	0.01%	0.0100	86.137

Display filter: Enter a display filter ...

So, if you see the majority of your data (and we can see we have 1.60%) is in the range 1280-2559, that means most of your data is being sent and received properly. If you have less of a percentage here and you have a large percentage of packets that are in a smaller packet length, then you might have a problem.

> Note that you will see a large percentage in the very small packet length there because you'll have a lot of the AKS and control packets.

I/O graph

Let's look at the I/O graph now. Click on **Statistics** and go to **I/O Graph**:

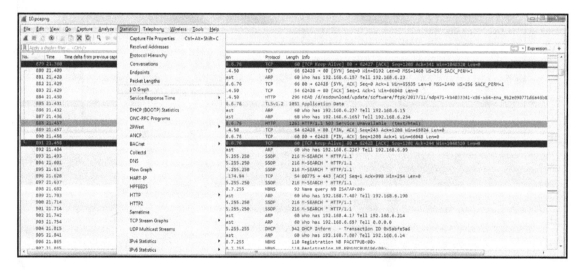

I/O graphs is a great graphical representation of the amount of data that's flowing through the packet capture, from the beginning to the end. I clicked on **Start** on the capture. Then, I went to my browser and opened it and typed in `http://www.pbs.org/`, and then pressed *Enter*. Then we saw the spike of traffic as it went and got that data from the content servers for PBS. And then that communication ended and I stopped the capture:

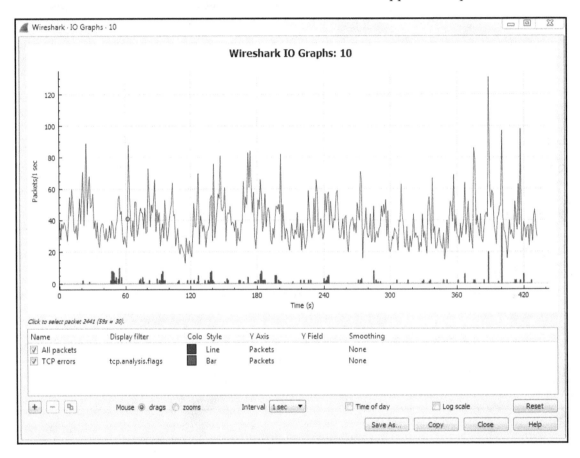

If you have a consistent file transfer that's occurring, for example, and it's at a low I/O due to some problem, you can see that in the I/O graph, and we'll get into the details of customizing this later on in a future chapter.

Load distribution

A useful statistic for HTTP is if you go down to the HTTP section by navigating to **Statistics | HTTP | Load Distribution**:

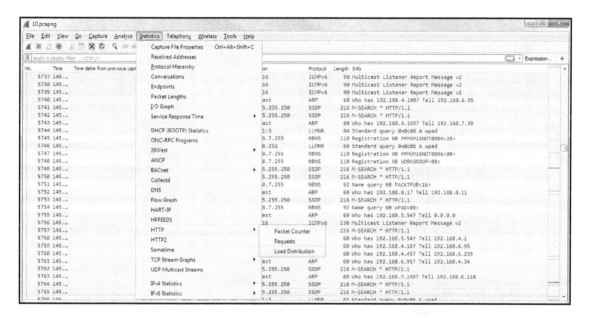

This will give you a view of all of the servers that have served up **HTTP** traffic in the packet capture:

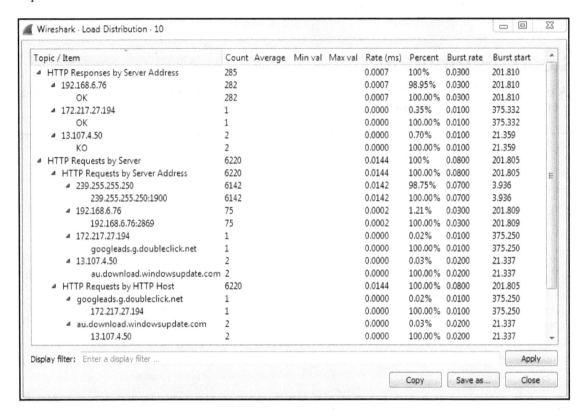

You will be able to see how many servers were served up, so if you're trying to diagnose a problem with a content distribution network, you could see that under the **Topic /** **Item** column; or if you're trying to find out which server is being loaded the most, because it has the most connections to it, you can see that under the **Count** column. You can add a filter in **Display filter** as well, if required.

DNS statistics

If you need to do a lot of analysis of DNS traffic, you can go to **Statistics** and then to **DNS**:

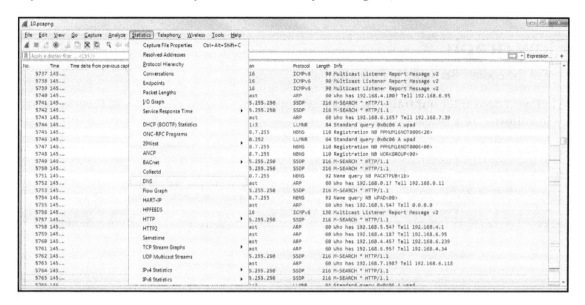

It'll give you a nice listing of all of the different codes that are returned in **DNS**:

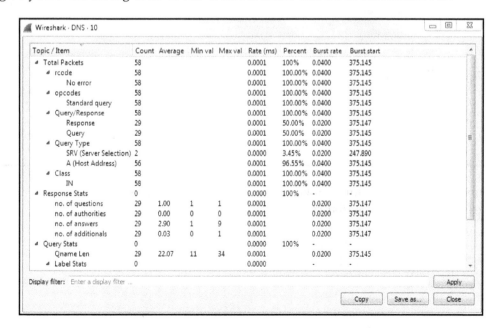

If you're having problems with your responses or you're getting a lot of errors when you're not expecting to receive any, you can view that easily instead of filtering by **DNS** traffic and going packet by packet through all of them.

Flow graph

One of the coolest **Statistics** is flow graph.

What we'll do is filter our traffic by this TCP flow. So we'll right-click on a TCP packet and go to **Follow** | **TCP Stream**:

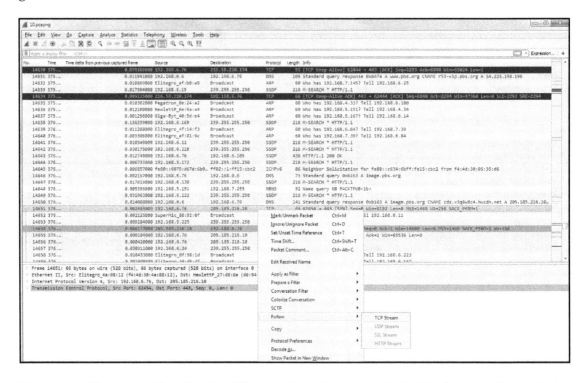

Now, we will create a filter for that. We've snipped out just a little bit of our traffic, and what we'll do next is go to **Statistics** and click on **Flow Graph**:

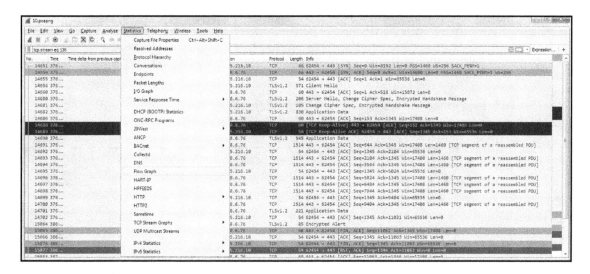

By default, it shows all of the packets in our packet capture:

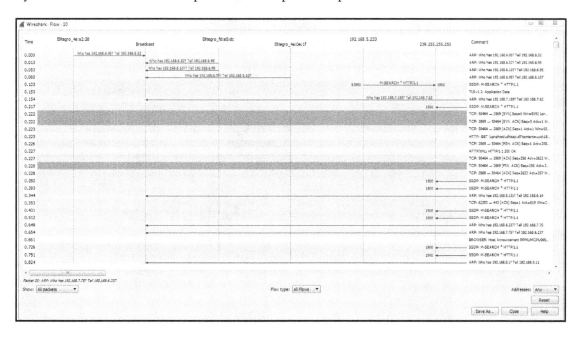

As we scroll through, you'll see all of the flows, all of the TCP handshakes-that is, the SYN, SYN, ACK, and ACK-and all of the requests and responses. Everything is in here, in a nice graphical format that you can use to diagnose your traffic. This is very useful for SIP traffic, for example.

What we can do is make the change in the bottom left where it says **Show**, and select **Displayed packets** here:

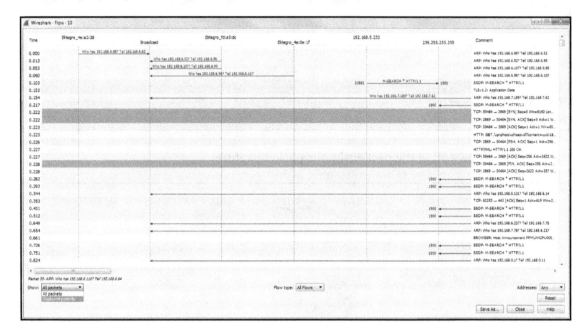

This change will be made for the filter that we have behind the screen. You can see the communication going from my computer to the server as I request for data such as a SYN:

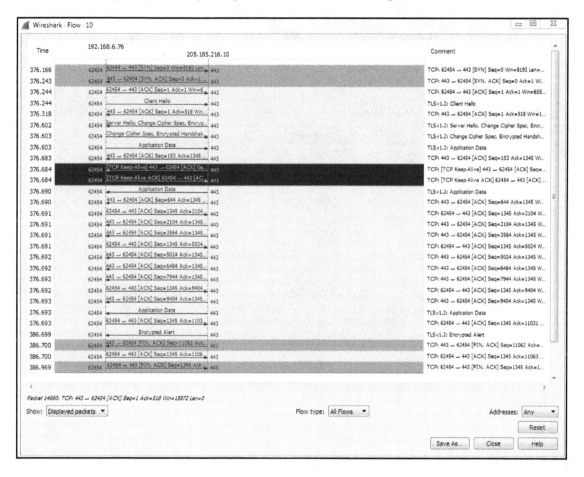

It's a very useful way of graphically representing the data flows that occur within your packet capture.

Expert system usage

In this section, we'll take a look at the expert system in Wireshark, which is a great feature that not many people know about, and it allows you to easily find problems in a packet capture.

You can follow along with the capture that I'll use by downloading the same one off of the Wireshark website. There's a great section of their wiki called **SampleCaptures** that allows you to download captures that have been submitted by the community:

What we'll do is search for errors, and the first one that comes up is `cmp-in-http-with-errors-in-cmp-protocol.pcap.gz`. If you download that and extract it, you can open up the `pcap` file and follow along:

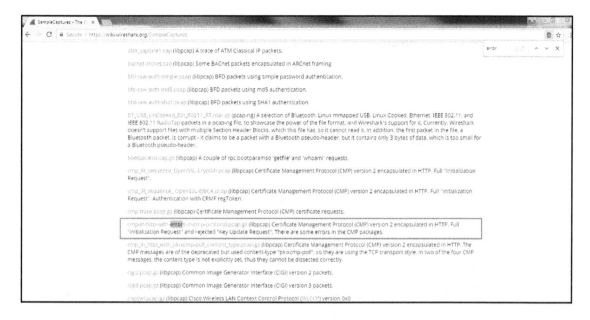

Here, we have our `pcap` file from the wiki, and there are two ways to get to the expert system. The first way is from the **Analyze** menu. We click on **Analyze**, and go to **Expert Information**:

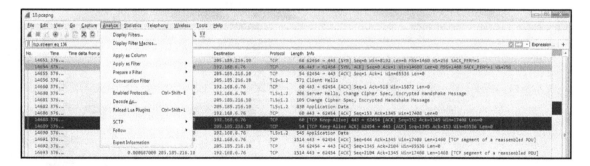

The second method is in the bottom left. As you can see in the following screenshot, it is a yellow colored icon:

This icon color will change based on the errors that are available in the packet capture. If this is yellow, the highest problem in this packet capture is a warning. If this is red, the highest problem in this packet capture is an error. If it's blue, it's chat information or informational. If you click on that, it will bring up the expert information. As you can see, there is a listing of all of the problems that Wireshark has automatically found in the packet capture:

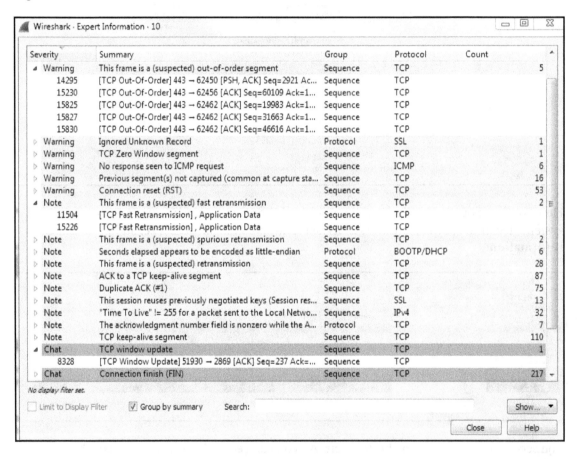

The **chat** ones are usually just generic informational data that you don't need to see, so we can minimize them. On your packet capture, if you have **warnings** or errors, these are things you want to take care of. The errors are the most critical. These are problems in the actual TCP communications or there's a problem in the packet in some way. There's a malformed issue or a CRC failure. Warnings are usually application problems, weird responses, and spooky stuff like that. Wireshark notices that and will alert you to it with this expert information window. Not only does it show you the type of errors and warnings it finds, but it tells you what packet number it's available on, and you can click on it. If we click on packet number `14295`, it will jump in the packet list down to `14295`, select it, and then show us what it's talking about. If we scroll down and look, we can see that it's under the **Transmission Control Protocol**:

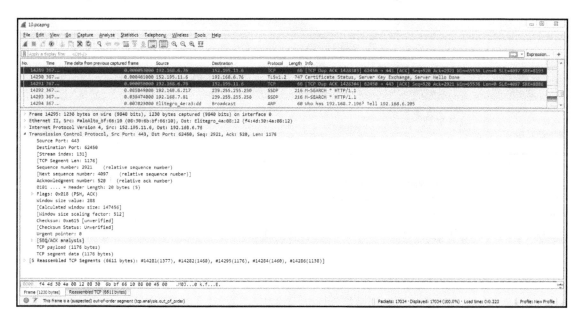

For another example we can take a look at on the **SampleCaptures**, if you search for x400, the first one that we'll get is this x400-ping-refuse.pcap:

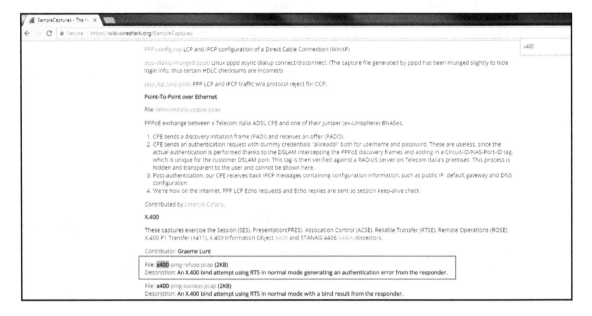

You can download the highlighted file as shown in the preceding screenshot and open it with Wireshark. If we go into expert information, you can see that there is a warning on packet 10. There's a Connection reset (RST), so we can click on that. It takes us to packet 10, and highlights exactly what the problem is:

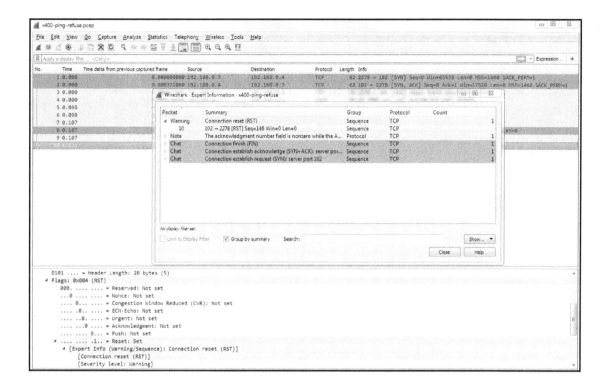

Summary

In this chapter, we've gone over the basics of TCP/IP and how a packet is built. We've seen how to use the time information in a packet capture to our advantage, and how to add time columns and change the time settings. Furthermore, we've seen how to view different statistics in Wireshark on our packet capture to help get an idea of what the capture contains and how to use the expert system, which is a nice hidden feature in Wireshark to show us exactly what the problems are in the packet capture that Wireshark knows about.

In Chapter 6, *Introductory Analysis*, we'll see an introduction to analysis of different protocols.

Summary

6
Introductory Analysis

The following are the topics that we'll cover in this chapter:

- DNS analysis
- ARP analysis
- IPv4 and v6 analysis
- ICMP analysis

We will see how each one of these is useful. Let's get started!!

DNS analysis

Let's take a look at how DNS works at a basic level and how to do common tasks with DNS such as look in a Wireshark capture.

We will start by flushing the DNS cache on the computer, which will clear out any of the cached entries on the device so that if we try to resolve any of them, it will have to go get a new resolution from the servers out there on the internet. For this, we will enter the following command:

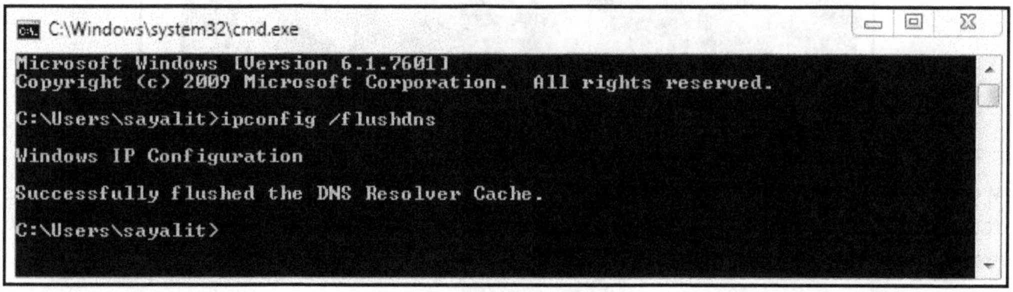

```
C:\Windows\system32\cmd.exe

Microsoft Windows [Version 6.1.7601]
Copyright (c) 2009 Microsoft Corporation.  All rights reserved.

C:\Users\sayalit>ipconfig /flushdns

Windows IP Configuration

Successfully flushed the DNS Resolver Cache.

C:\Users\sayalit>
```

Since we've cleared that out, let's do a standard resolution.

Before we do a resolution, what DNS does is resolve domain names and different records of these domain names to IP addresses. That's its primary purpose.

DNS is used for all sorts of things, some of which are listed here:

- Browsing the internet in any fashion, such as with the web
- If you're trying to resolve FTP servers or game servers
- If you're trying to run an active directory on a domain `011` into a local network
- If you're trying to run VMware

All sorts of different services out there use DNS; there's a common mantra in IT that even when you think it's not DNS, it's usually DNS whenever there's a problem. Let's take a look at what a normal DNS resolution looks like.

For that we will type the following command in Command Prompt:

```
nslookup wireshark.org 8.8.8.8
```

We'll force that query to go out to Google directly. If I press *Enter*, we get the following output:

So, in this result, you see that we have the server that responded to our query, `8.8.8.8`, and we can also see the DNS name for that server address. You can also see the answers that it has for that device for that server, and you can see that we have both IPv6 and IPv4 addresses.

The IPv6 addresses, as you can see, have a very different format from IPv4.

> If you're unfamiliar with IPv6, take a look at the other books that are available from Packt Publishing regarding IPv6.

So what we want to do is take a look at this from a packet level. For this, we'll flush the DNS cache again:

```
ipconfig /flushdns
```

Next, we'll start a capture on the local interface, as shown in the following screenshot:

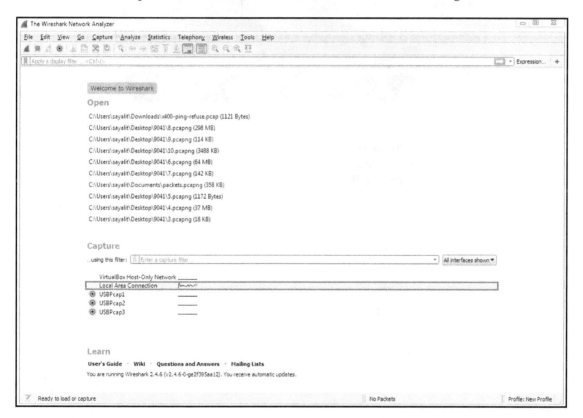

While the capture's running, let's go ahead and enter the same command we used earlier:

```
nslookup wireshark.org 8.8.8.8
```

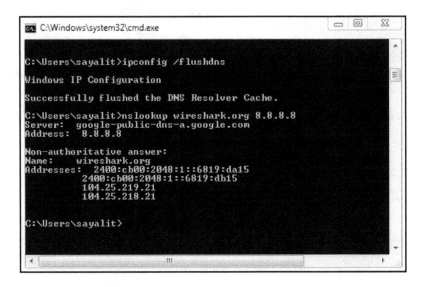

Next, we will stop the capture. If you scroll down, you can see that there's a whole bunch of DNS as well as some other protocols. So what we'll do is, we'll use a filter and simply type in dns in the display filter. That will get rid of any other junk that we don't want to see:

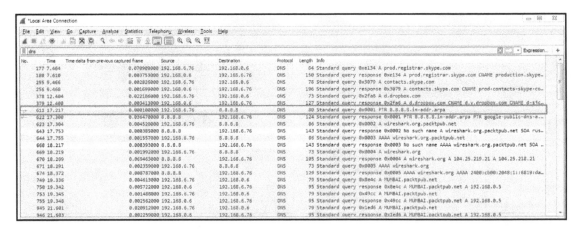

You can see the first request. So we have some other DNS requests that occurred by the system. We'll take a look at the first query highlighted in the preceding screenshot.

You can see in the query that it's asking for 8.8.8.8, and so it's actually asking for the domain name of the domain server. You can also see that it has a Transaction ID of 1.

If we look at the next query that happened, wireshark.org, you'll see that it has a Transaction ID of 2:

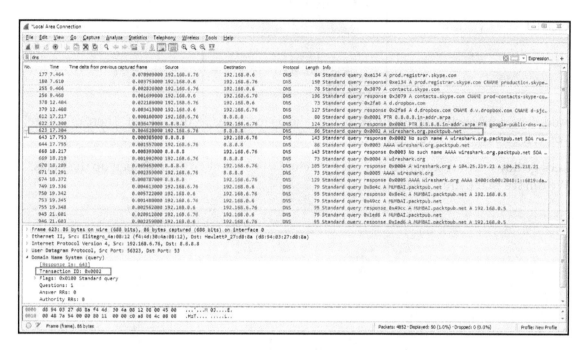

If we look at the packets in response to the query, they will have a matching transaction ID.

So a good example as to why this is useful is if you see a number of responses, or queries even, that are going around on your network and showing up in your packet capture with the same transaction ID over and over again, you may have a loop in your network; that could point to a problem with spanning tree, for example.

It's also just useful in general for us to be able to determine which packets are matched up from query to response. As you can see, Wireshark is automatically showing us the related packets between the two, as we've already mentioned in the previous sections.

We also have a line that tells us within the DNS section of the packet details that it's a response in packet 622:

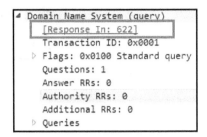

And if we double-click on **Request In**, it'll take us to the respective packet. Then, of course, you can go back and forth between the two. So, the transaction ID is very useful.

Let's go down to our second query for the actual wireshark.org, and open up the **Flags**. We can see that there's a flag turned on:

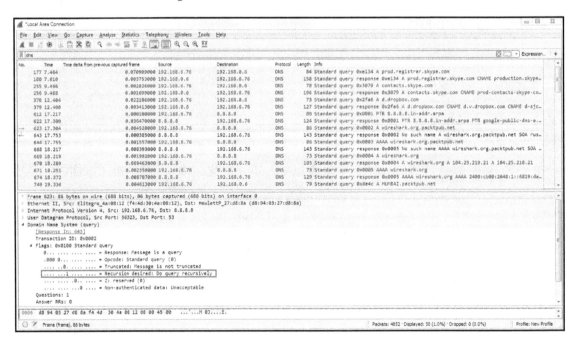

We have a 1 bit enabled, where it says **Recursion desired: Do query recursively**. This means that the query is requesting the server to ask other servers in case it does not have the answer to our query. A DNS server could have additional pointers configured in it, or additional forwarders set up on it, to go look for the answer to a DNS query. So, this query flag is saying "yes, go ahead and do that for us".

Let's go down to packet 669, where the system requests the wireshark.org A record from Google. We can see the response on packet 670. Then, we dig into the **Flags**, which gives the highlighted response in the following screenshot:

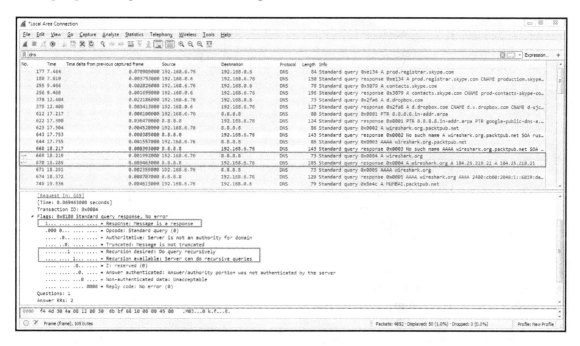

Bit 1 is enabled, so the highlighted part is a response message. It has Recursion desired and Recursion available as enabled.

Now, we'll scroll down and see that we have **Answer RRs: 2**, so we have two responses to this:

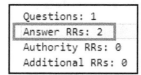

We'll see the **Queries** and when we go down to **Answers**, we'll see that we have two records for `wireshark.org`:

```
⊿ Queries
    ⊿ wireshark.org: type A, class IN
          Name: wireshark.org
          [Name Length: 13]
          [Label Count: 2]
          Type: A (Host Address) (1)
          Class: IN (0x0001)
⊿ Answers
    ⊿ wireshark.org: type A, class IN, addr 104.25.219.21
          Name: wireshark.org
          Type: A (Host Address) (1)
          Class: IN (0x0001)
          Time to live: 299
          Data length: 4
          Address: 104.25.219.21
    ⊿ wireshark.org: type A, class IN, addr 104.25.218.21
          Name: wireshark.org
          Type: A (Host Address) (1)
          Class: IN (0x0001)
          Time to live: 299
          Data length: 4
          Address: 104.25.218.21
```

Thus, it looks like `wireshark.org` is using some sort of load-balancing system because we have an A record of `104.25.219.21` and `104.25.218.21`. If you go to `https://www.wireshark.org/`, it will take you to either of these two addresses. You'll also see that there's a **Time to live value**, where you have a TTL of `299`, which is the number of seconds that this record is to be kept on my system before it requests a new record. The TTL value is very short here, and most likely this is something to do with the fact that they are using a load balancer. If there's any sort of change in the address, it wants us to get an update as quickly as possible. It doesn't want us to keep a cached version of bad IP addresses for a very long period of time. For example, many defaults are 8 hours or 24 hours; so, converted to seconds, you might see 86,400 seconds or so.

An example for DNS request failure

Let's take a look at a DNS request that will fail.

For this, we'll create some sort of random domain and put in gibberish, `jhadgug384r8.com`, as shown in the following screenshot:

```
C:\Users\sayalit>nslookup jhadgug384r8.com 8.8.8.8
```

Next, you need to start a new capture. Go ahead and press *Enter*:

```
C:\Windows\system32\cmd.exe

C:\Users\sayalit>nslookup jhadgug384r8.com 8.8.8.8
Server:  google-public-dns-a.google.com
Address:  8.8.8.8

*** google-public-dns-a.google.com can't find jhadgug384r8.com: Non-existent dom
ain

C:\Users\sayalit>
```

Now, you need to stop the capture.

We can see we have the requests, like before:

No.	Time	Time delta from previous captured frame	Source	Destination	Protocol	Length	Info
252	7.400	0.014623000	192.168.6.76	192.168.0.6	DNS	73	Standard query 0x5800 A web.skype.com
253	7.410	0.001601000	192.168.0.6	192.168.6.76	DNS	201	Standard query response 0x5800 A web.skype.com CNAME webclientshellserver-prod.trafficmanager.net
290	7.877	0.024495000	192.168.6.76	8.8.8.8	DNS	80	Standard query 0x0001 PTR 8.8.8.8.in-addr.arpa
292	7.950	0.013006000	8.8.8.8	192.168.6.76	DNS	124	Standard query response 0x0001 PTR 8.8.8.8.in-addr.arpa PTR google-public-dns-a.google.com
293	7.953	0.003524000	192.168.6.76	8.8.8.8	DNS	89	Standard query 0x0002 A jhadgug384r8.com.packtpub.net
312	8.444	0.008007000	8.8.8.8	192.168.6.76	DNS	146	Standard query response 0x0002 No such name A jhadgug384r8.com.packtpub.net SOA rush.easydns.com
313	8.448	0.002045000	192.168.6.76	8.8.8.8	DNS	89	Standard query 0x0003 AAAA jhadgug384r8.com.packtpub.net
334	8.952	0.038570000	8.8.8.8	192.168.6.76	DNS	146	Standard query response 0x0003 No such name AAAA jhadgug384r8.com.packtpub.net SOA rush.easydns.com
336	8.954	0.001466000	192.168.6.76	8.8.8.8	DNS	76	Standard query 0x0004 A jhadgug384r8.com
341	9.209	0.010506000	8.8.8.8	192.168.6.76	DNS	149	Standard query response 0x0004 No such name A jhadgug384r8.com SOA a.gtld-servers.net
342	9.211	0.001981000	192.168.6.76	8.8.8.8	DNS	76	Standard query 0x0005 AAAA jhadgug384r8.com
351	9.440	0.001103000	8.8.8.8	192.168.6.76	DNS	149	Standard query response 0x0005 No such name AAAA jhadgug384r8.com SOA a.gtld-servers.net

We have the `8.8.8.8` domain name, and we have the actual request. Just like with `wireshark.org`, you can see that we've received an A record as well as a quad A record. This is because the system is requesting both IPv4 and IPv6. You saw that in the command-line output when we did `wireshark.org`.

So we have an A record, which is the IPv4, and a quad A record, which is IPv6. But you can see that we have some sort of gibberish domain name, which obviously responded back in that command line, saying that it couldn't find a result. So we sent out a query, asking for `jhadgug384r8.com`, and we received a response from the DNS servers at Google, saying `No such name`. If we go to the **Flags** and take a look, we have `Message is a response`, `No such name`, `Reply code: No such name`.

And if you remember from the filtering section, we can right-click on any flag and apply a filter by selecting **Apply as Filter** based on the options available, as shown in the following screenshot:

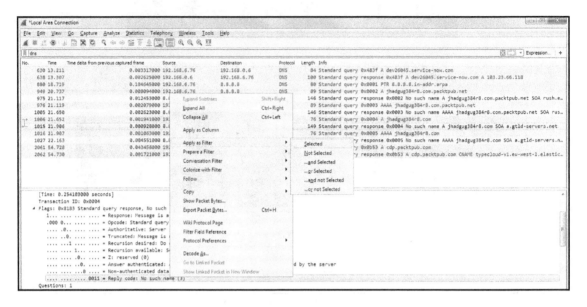

So we just filtered our display filter based on No such name as the error code in DNS. This is a great filter you could apply if you're doing a packet capture where there's some sort of connectivity problem. You can look for failed DNS queries.

ARP analysis

In this section, we'll take a look at how ARP works, resolve addresses from IP to MAC, and also see what ARP issues look like in Wireshark. So what ARP does is resolve the IP addresses, which are layer 3 addresses, to MAC addresses, which are layer 2 addresses—these are addresses that are used on our local Ethernet bus. We need this information in order to construct a frame which encapsulates a packet, so we can send it on to the wire. When a user or an application requests data from a specific IP address on layer 3, our system has to figure out what that MAC address is, if it doesn't already have it in its cache. We can check what MAC addresses our system already knows about in its ARP cache. Just like DNS had a cache of locally known information, ARP is also locally cached.

So what we can do is type the following in a Windows machine:

```
arp -a
```

If you press *Enter*, you'll get a list of all the known IP addresses in layer 3 matched up with the physical addresses, which are the MAC addresses on layer 2:

```
Command Prompt

Interface: 192.168.159.1 --- 0x11
  Internet Address      Physical Address      Type
  192.168.159.255       ff-ff-ff-ff-ff-ff     static
  224.0.0.2             01-00-5e-00-00-02     static
  224.0.0.22            01-00-5e-00-00-16     static
  224.0.0.251           01-00-5e-00-00-fb     static
  224.0.0.252           01-00-5e-00-00-fc     static
  224.0.0.253           01-00-5e-00-00-fd     static
  239.2.0.252           01-00-5e-02-00-fc     static
  239.255.250.250       01-00-5e-7f-fa-fa     static
  239.255.255.250       01-00-5e-7f-ff-fa     static
  255.255.255.255       ff-ff-ff-ff-ff-ff     static

Interface: 192.168.139.1 --- 0x12
  Internet Address      Physical Address      Type
  192.168.139.255       ff-ff-ff-ff-ff-ff     static
  224.0.0.2             01-00-5e-00-00-02     static
  224.0.0.22            01-00-5e-00-00-16     static
  224.0.0.251           01-00-5e-00-00-fb     static
  224.0.0.252           01-00-5e-00-00-fc     static
  224.0.0.253           01-00-5e-00-00-fd     static
  239.2.0.252           01-00-5e-02-00-fc     static
  239.255.250.250       01-00-5e-7f-fa-fa     static
  239.255.255.250       01-00-5e-7f-ff-fa     static
```

In the preceding screenshot, what you see in the third column is, as it says, `Type static`.

All of these addresses are the ones that my system knows about because they are coded into the operating system at this moment. So, these addresses are multicast addresses that the system knows about by default. It also knows some VMware interface, IP and MAC addresses, and a number of things that already exist.

If we scroll up, we can see that there are some dynamically learned addresses:

```
Command Prompt

C:\Users\Andrew>arp -a

Interface: 192.168.77.159 --- 0xa
  Internet Address      Physical Address      Type
  192.168.77.1          00-7f-28-e7-bf-47     dynamic
  192.168.77.89         ec-08-6b-f9-ea-c6     dynamic
  192.168.77.96         f4-81-39-92-ab-18     dynamic
  192.168.77.97         00-1f-33-eb-0e-3e     dynamic
  192.168.77.98         b8-27-eb-24-9f-84     dynamic
  192.168.77.99         a4-2b-b0-aa-c0-50     dynamic
  192.168.77.153        ac-5f-3e-a0-59-84     dynamic
  192.168.77.161        98-5f-d3-45-6a-aa     dynamic
  192.168.77.162        0c-47-c9-21-89-58     dynamic
  192.168.77.255        ff-ff-ff-ff-ff-ff     static
  224.0.0.2             01-00-5e-00-00-02     static
  224.0.0.22            01-00-5e-00-00-16     static
  224.0.0.251           01-00-5e-00-00-fb     static
  224.0.0.252           01-00-5e-00-00-fc     static
  224.0.0.253           01-00-5e-00-00-fd     static
  239.2.0.252           01-00-5e-02-00-fc     static
  239.255.250.250       01-00-5e-7f-fa-fa     static
  239.255.255.250       01-00-5e-7f-ff-fa     static
  255.255.255.255       ff-ff-ff-ff-ff-ff     static
```

Next, you can see some statically known information—some multicast for the interface
192.168.77.159.

You can see the primary interface, where we have the static information that's known for
that interface, and we have dynamically learned addresses. The addresses under Physical
Address are all MAC addresses that have been discovered for specific IPs and are cached
for a specific period of time. Then, if I need to access that device again on layer 2, it will do
an ARP request again.

Now, what we'll do is take a packet capture of a normal, good ARP request.

We'll start the capture and ping a known good address on IPv4:

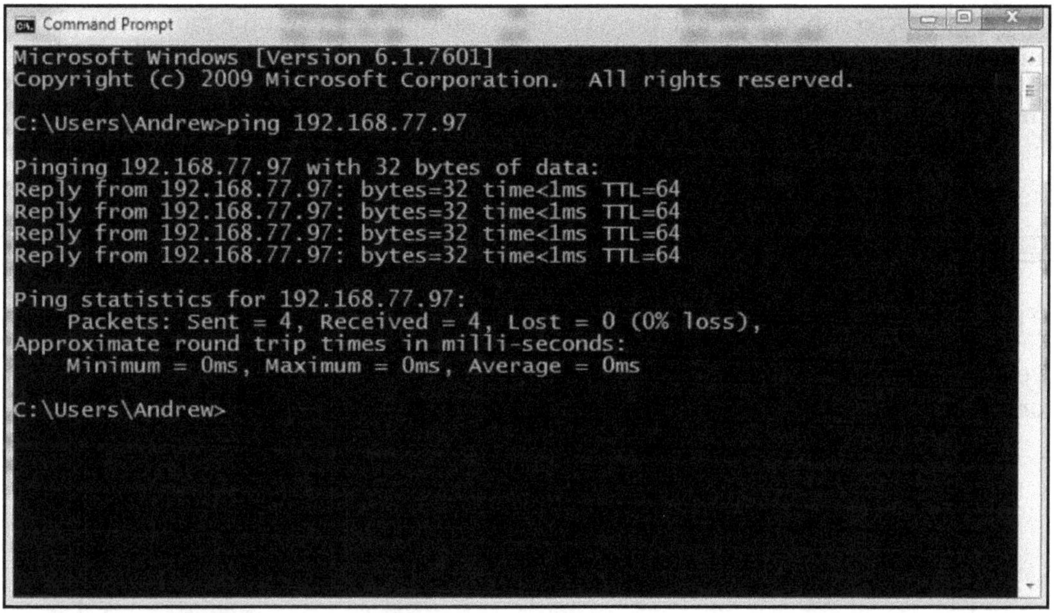

As we can see, we received a number of replies. The system has sent four pings using **Internet Control Message Protocol** (**ICMP**) to this device, and received a response to all four of them. Before it was able to do this, it had to figure out what the local physical address is of that device, what the MAC address was, before it could even do the first response, or first request.

Next, we'll stop our capture and do a filter for `arp`:

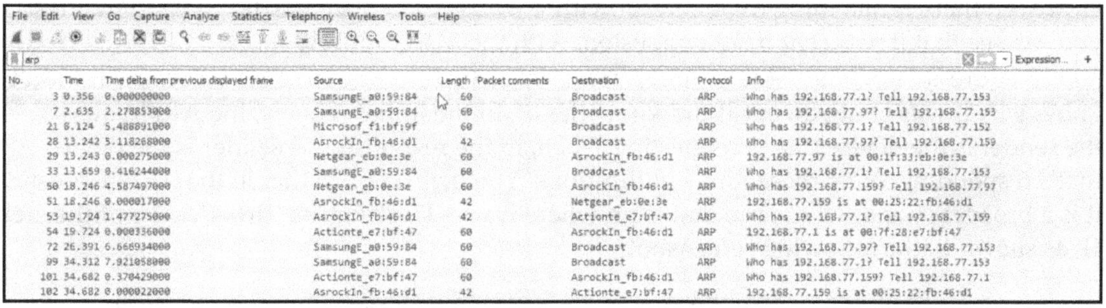

That way, we only see our ARP traffic and we'll skip some of the information. There are some other devices on the network that are also trying to do ARP requests, but if we check the preceding screenshot, we have the `AsrockIn` interface, which is the motherboard of the computer we are using.

In the **Info** column, you can see `Who has 192.168.77.97? Tell 192.168.77.159`. That's the IP address of the system we are using right now. We get a response saying, `192.168.77.97 is at 00:1f:33:eb:0e:3e` MAC address:

```
Netgear_eb:0e:3e          60                AsrockIn_fb:46:d1    ARP       192.168.77.97 is at 00:1f:33:eb:0e:3e
```

The preceding screenshot shows a Netgear NAS device that we have on our network.

If we go into the ARP information in the packet details pane, we can see the same information as shown in the preceding screenshot, but in summary form:

```
▷ Frame 29: 60 bytes on wire (480 bits), 60 bytes captured (480 bits) on interface 0
▷ Ethernet II, Src: Netgear_eb:0e:3e (00:1f:33:eb:0e:3e), Dst: AsrockIn_fb:46:d1 (00:25:22:fb:46:d1)
◢ Address Resolution Protocol (reply)
     Hardware type: Ethernet (1)
     Protocol type: IPv4 (0x0800)
     Hardware size: 6
     Protocol size: 4
     Opcode: reply (2)
     Sender MAC address: Netgear_eb:0e:3e (00:1f:33:eb:0e:3e)
     Sender IP address: 192.168.77.97
     Target MAC address: AsrockIn_fb:46:d1 (00:25:22:fb:46:d1)
     Target IP address: 192.168.77.159
```

It's an Ethernet response. The sender MAC in the frame is from Netgear. It's being sent from Netgear's IPv4 address to the initial requester and then to our system with the IP `192.168.77.159`.

However, there is one thing to be careful with. I we look at where we did the ARP request, you can see that it was sent from our system with the MAC address `00:25:22:fb:46:d1` and IP address `192.168.77.159`, but it was sent to `00.00.00_00:00:00` because it didn't know who it was going to, but it did know the IP. In the initial request, the ARP request, the sender is the device that is requesting. Then, in the response, the sender is the device that's responding. Note these kinds of flip-flops. You'll also notice that in the ARP request, it is a broadcast, and we can see that from where it says **Destination: Broadcast** on **Ethernet II**, as shown in the following screenshot:

```
Ethernet II, Src: AsrockIn_fb:46:d1 (00:25:22:fb:46:d1), Dst: Broadcast (ff:ff:ff:ff:ff:ff)
▷ Destination: Broadcast (ff:ff:ff:ff:ff:ff)
▷ Source: AsrockIn_fb:46:d1 (00:25:22:fb:46:d1)
```

It's going to a broadcast address. So when our system is trying to find on the local Ethernet bus, the MAC address of the device that has the IP shown on layer 3, it doesn't know who to talk to for that so it sends out a broadcast to everyone. Then, the device that happens to know that information receives it (because everybody has a copy) and responds. Similarly, with DNS, if you see a bunch of repeats here, especially response frames, that are looping over and over again—they just keep showing up in your capture constantly—that could again point to a loop in your network. You shouldn't be seeing these over and over and over again.

An example for ARP request failure

Let's take a look at an example where an ARP request fails.

What we'll do is start a new capture, and this time we will ping an address that does not exist on my network:

```
C:\Users\Andrew>ping 192.168.77.124

Pinging 192.168.77.124 with 32 bytes of data:
Reply from 192.168.77.159: Destination host unreachable.
Reply from 192.168.77.159: Destination host unreachable.
Reply from 192.168.77.159: Destination host unreachable.
Reply from 192.168.77.159: Destination host unreachable.

Ping statistics for 192.168.77.124:
    Packets: Sent = 4, Received = 4, Lost = 0 (0% loss),
```

We will see `Destination host unreachable` show up. We will do that a couple of times. Since we are trying a ping, it's going to do that four times on a Windows system. If you're on Linux or macOS, it'll probably do that nonstop, depending on what you use.

Go ahead and stop that capture; we have plenty of information here:

No.	Time	Time delta from previous displayed frame	Source	Length	Packet comments	Destination	Protocol	Info
5	3.489	0.000000000	SamsungE_a0:59:84	60		Broadcast	ARP	Who has 192.168.77.1? Tell 192.168.77.153
13	7.049	3.559666000	SamsungE_a0:59:84	60		Broadcast	ARP	Who has 192.168.77.97? Tell 192.168.77.153
20	9.158	2.109185000	Actionte_e7:bf:47	60		Broadcast	ARP	Who has 192.168.77.10? Tell 192.168.77.1
22	10.158	0.999940000	Actionte_e7:bf:47	60		Broadcast	ARP	Who has 192.168.77.10? Tell 192.168.77.1
24	11.158	0.999983000	Actionte_e7:bf:47	60		Broadcast	ARP	Who has 192.168.77.10? Tell 192.168.77.1
34	17.694	6.536626000	AsrockIn_fb:46:d1	42		Broadcast	ARP	Who has 192.168.77.124? Tell 192.168.77.159
37	18.639	0.944542000	AsrockIn_fb:46:d1	42		Broadcast	ARP	Who has 192.168.77.124? Tell 192.168.77.159
43	19.639	1.000033000	AsrockIn_fb:46:d1	42		Broadcast	ARP	Who has 192.168.77.124? Tell 192.168.77.159
45	20.640	1.001061000	AsrockIn_fb:46:d1	42		Broadcast	ARP	Who has 192.168.77.124? Tell 192.168.77.159
47	20.853	0.212639000	SamsungE_a0:59:84	60		Broadcast	ARP	Who has 192.168.77.1? Tell 192.168.77.153
50	21.638	0.785383000	AsrockIn_fb:46:d1	42		Broadcast	ARP	Who has 192.168.77.124? Tell 192.168.77.159
51	22.138	0.500021000	AsrockIn_fb:46:d1	42		Actionte_e7:bf:47	ARP	Who has 192.168.77.1? Tell 192.168.77.159
52	22.138	0.000510000	Actionte_e7:bf:47	60		AsrockIn_fb:46:d1	ARP	192.168.77.1 is at 00:7f:28:e7:bf:47
54	22.638	0.499782000	AsrockIn_fb:46:d1	42		Broadcast	ARP	Who has 192.168.77.124? Tell 192.168.77.159
128	23.639	1.000777000	AsrockIn_fb:46:d1	42		Broadcast	ARP	Who has 192.168.77.124? Tell 192.168.77.159
130	24.415	0.776341000	SamsungE_a0:59:84	60		Broadcast	ARP	Who has 192.168.77.97? Tell 192.168.77.153
131	24.638	0.222095000	AsrockIn_fb:46:d1	42		Broadcast	ARP	Who has 192.168.77.124? Tell 192.168.77.159
132	25.268	0.629300000	Actionte_e7:bf:47	60		Broadcast	ARP	Who has 192.168.77.10? Tell 192.168.77.1

We can see the request highlighted in the preceding screenshot: Who has
192.168.77.124?

We can see the target down in the **Address Resolution Protocol (request)** option. And we
just keep requesting it over and over and over again. We're trying desperately to find who
has that MAC address for IP 192.168.77.124, but it doesn't exist. It just keeps trying and
trying, and it keeps timing out and failing; and that will show up in our results, as you saw
in the command line, as unreachable.

IPv4 and IPv6 analysis

We'll now take a look at the differences between IPv4 and IPv6 and learn about issues and
features such as the fragmentation of these packets, broadcast storms, and flags within the
IPv4 and the IPv6 header.

What we have is some data from a packet capture going to a website which was encrypted,
so that's why we see a lot of TLS in the protocol information:

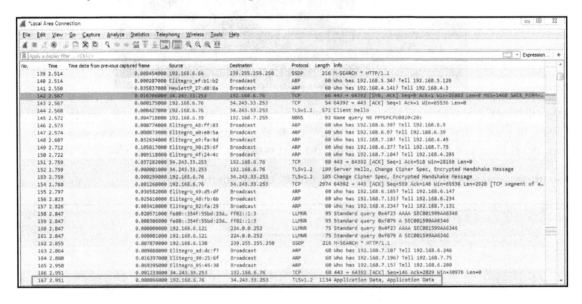

And we see that we have `Application Data` in the **Info** column, which is all of the encrypted data transmitting back and forth to the web server. Go to IPv4 in the packet details, expand that, and we can take a look at the information in the IPv4 header:

```
▲ Internet Protocol Version 4, Src: 192.168.6.76, Dst: 34.243.33.253
     0100 .... = Version: 4
     .... 0101 = Header Length: 20 bytes (5)
   ▷ Differentiated Services Field: 0x00 (DSCP: CS0, ECN: Not-ECT)
     Total Length: 1120
     Identification: 0x2ddf (11743)
   ▷ Flags: 0x4000, Don't fragment
     Time to live: 128
     Protocol: TCP (6)
     Header checksum: 0x0000 [validation disabled]
     [Header checksum status: Unverified]
     Source: 192.168.6.76
     Destination: 34.243.33.253
```

We can see that, right after **Internet Protocol Version 4**, it's saying that it's `Version: 4`; otherwise, it will show `Version 6`. It also has the `Header Length`, which is the number of bytes in the header. Sometimes, the `Header Length` can fluctuate, so it defines how big that header is so that the application knows where the differentiating point is between the header and the actual data in the packet. We also have some DSCP information for quality of service purposes and the total length of the packet. If you're familiar with MTU, such as setting the MTU of the interface of a router or a computer, that's where this comes into play. `Total Length` is the total size of that packet. If the total size of a packet is too large, it will fragment.

Also, we can see that it says `Fragment offset` inside the **Flags**. If we expand our **Flags**, we can see that we have fragmentation settings in the **Flags**. So we have `More fragments` that are coming, or `Don't fragment`. The `Fragment offset` tells the IP stack where to pick up with the additional data that's coming so that it can combine it into one large packet:

```
▲ Flags: 0x4000, Don't fragment
     0... .... .... .... = Reserved bit: Not set
     .1.. .... .... .... = Don't fragment: Set
     ..0. .... .... .... = More fragments: Not set
     ...0 0000 0000 0000 = Fragment offset: 0
```

We also have a TTL, which usually is some sort of default number, such as 24 or 60, or something like 128. As it hops between the different routers throughout the internet or your local network-wherever it happens to be destined to go-it will decrement 1 as it goes through each device. If it reaches 1 when it's received on a router-if it's a TTL of 1-then that router will discard that packet. If a host machine receives a packet with a TTL of 1, it will process it because it doesn't have to actually route it somewhere. The TTL prevents packets from looping around forever throughout a network. If it loops through 60 devices, it will then be discarded; so, it won't get stuck there forever.

We also have a protocol definition: is it a TCP packet, a UDP packet, or some other type of protocol. And we have some checksum information to ensure that the header has not been manipulated in any way. Notice that it's not a checksum of the entire block of data, such as the FCS at the end of a frame which encapsulates this, but it's the header checksum to make sure that the header itself is not manipulated. Then, of course, we have the source and destination addresses that we're coming from and going to, respectively:

```
Time to live: 128
Protocol: TCP (6)
Header checksum: 0x0000 [validation disabled]
[Header checksum status: Unverified]
Source: 192.168.6.76
Destination: 34.243.33.253
```

Let's take a look at a packet that's been fragmented:

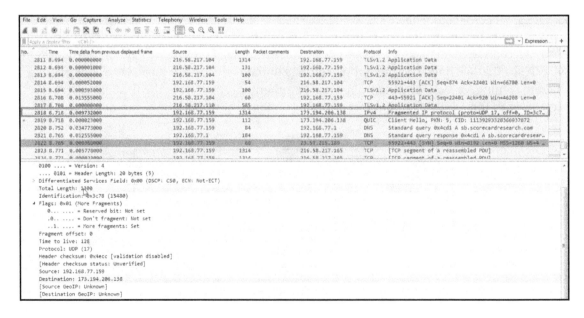

We can actually see that we have an IPv4 packet and it says `Fragmented IP protocol` in Wireshark here; it knows that it's a fragmented packet. We see in the packet details that it has a length of `1300` bytes, and in the flags it has the 1 bit turned on for more fragments. So it says there is a fragment coming up. There's an additional packet; that next packet is part of this, so we can combine them. We can see that `Protocol` is a UDP packet and that it was generated with a TTL of `128`. Since the source was our local machine and it's going out to the internet, we know that this is the generated TTL because it's being recorded and captured on the device that is sending it. So, our system is actually defaulting the detail to `128`.

The next packet is a continuation of the preceding packet:

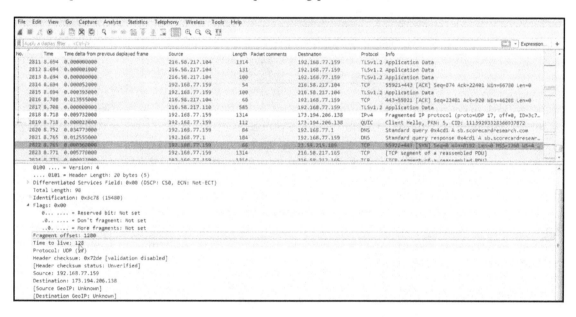

We can see that `Fragment offset` is `1280`, so it knows that it needs to be combined with that previously captured packet. You can see that, if you go down in the packet details, Wireshark actually combines them in this little information section in the details. If we expand **[2 IPv4 Fragments (1358 bytes): #2818(1280), #2819(78)]**, it says that there are two packets involved:

```
▲ [2 IPv4 Fragments (1358 bytes): #2818(1280), #2819(78)]
    [Frame: 2818, payload: 0-1279 (1280 bytes)]
    [Frame: 2819, payload: 1280-1357 (78 bytes)]
    [Fragment count: 2]
    [Reassembled IPv4 length: 1358]
```

We can select the first and second packets. Click on them and you can see the following information:

```
▲ Flags: 0x01 (More Fragments)
    0... .... = Reserved bit: Not set
    .0.. .... = Don't fragment: Not set
    ..1. .... = More fragments: Set
  Fragment offset: 0
  Time to live: 128
  Protocol: UDP (17)
  Header checksum: 0x4ecc [validation disabled]
  [Header checksum status: Unverified]
  Source: 192.168.77.159
  Destination: 173.194.206.138
  [Source GeoIP: Unknown]
  [Destination GeoIP: Unknown]
  Reassembled IPv4 in frame: 2819
```

So Wireshark is smart enough to know this. It takes a look at the header information and provides you with the details, as shown in the preceding screenshot, so that you don't have to do the math yourself. It even says how many fragmented packets there are within this one transmission and also provides additional information, such as the total length of the reassembled data. An application can define whether or not it's fragmented. So when an application wants to communicate, it will tell the stack whether to set the Don't fragment bit or not. Depending on the application and its requirements, it may say that it does not want its data to be fragmented. Maybe it's an encrypted packet and, if you fragment it, it will mess up the encryption. Since it doesn't want to have that information fragmented, it will turn the bit on which says Don't fragment so that the IP stacks on both sides know that they need not fragment the data. If you notice, the initial packet that was sent-that's going to be fragmented-has an identification of 3c78. If we look at the second packet in this series, we see that its identification is also 3c78.

Remember when we were talking about the ID field? It changes based on each conversation or each packet that's sent. If the identification is the same, that's an indicator that the packet is fragmented. This is how Wireshark is actually combining them and realizing they're part of a series, because the ID is the same but the data within it is different. It's not a duplicate, but just a continuation of a fragmented piece of data. Now, in an IPv6 packet, you'll see that the header is somewhat similar to IPv4:

It's actually a little bit simplified. It has `Payload length`; it tells you what kind of data is within it: is it TCP or UDP; it has a TTL (they call it `Hop limit`); and it also has a source and destination address. Remember that the addresses look different in IPv6 as it uses hexadecimal.

ICMP analysis

In this section, we'll take a look at how ICMP is useful to network engineers and what some of the different types of ICMP are and what they mean.

The first thing we will do is create some ICMP packets. For that we will create a ping request, which is a type of ICMP.

So, let's go ahead and start the capture, and we'll go ahead and ping Google again:

Each one of the replies is a series of ICMP requests and responses.

Stop the capture and we'll apply an `icmp` display filter:

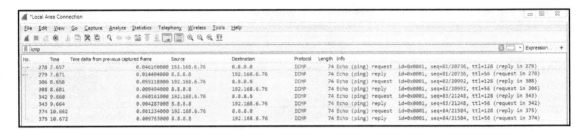

As shown in the preceding screenshot, these are all of my ICMP packets that have been sent and received, and we can see that we have multiple requests and replies. This coincides with the four replies that we saw in the command line.

If we dive into the header, by going down to the very bottom in the packet details—that is, to **Internet Control Message Protocol**, we can see that we have Type and Code:

```
▲ Internet Control Message Protocol
    Type: 8 (Echo (ping) request)
    Code: 0
    Checksum: 0x4d0a [correct]
    [Checksum Status: Good]
    Identifier (BE): 1 (0x0001)
    Identifier (LE): 256 (0x0100)
    Sequence number (BE): 81 (0x0051)
    Sequence number (LE): 20736 (0x5100)
    [Response frame: 279]
  ▷ Data (32 bytes)
```

Type and Code are the two important parts within ICMP. We see we have Type 8: (Echo (ping) request). Then, in the next packet we have Type: 0 (Echo (ping) reply). The Type and Code give us information about what's going on in the network. Now, this is a very simple example of requesting a poll of a device and a response from a device to see if it's alive. It's basically like ringing the doorbell of a device to see if it's there.

There are other types and codes within ICMP that are useful to our devices on our network and are also useful to us as engineers. A very simple example is an echo request and you can reply to see if something is accessible and if it's being allowed by the device, such as firewalls; they may be blocking the request, but it's a very useful thing. We also have other types and code we can look at.

If you go to the ICMP page on Wikipedia (`https://en.wikipedia.org/wiki/Internet_Control_Message_Protocol`), we can scroll down and take a look at all of these types and code. You can see that there are a number of different types, and there's code within each type.

You can see that there's type **8 - Echo Request**, used to ping; and type **0 - Echo Reply**. But we have these additional types that are here as well, such as type **3 - Destination Unreachable**:

A router may send an ICMP packet back to a device, letting it know that it's unable to access a certain network or host for it. You can see all the different types that it has to describe to the requesting device why it can't reach the specific resource that it's trying to get to.

The other common types are router advertisement and router solicitation. These allow hosts to request and receive a router to access a specific network. Additionally, we have type **11 - Time Exceeded**. Remember when we were talking about the TTL and how it starts at a certain number, such as `60` or `128` or something like that, and it will then decrement as it goes through the different routers throughout a network and across the internet? When a router receives a packet with a TTL of `1`, it will discard it and at the same time generate an ICMP packet and send it back to the original source device, letting it know that the TTL was exceeded and that it needed to discard it. It will do so with type `11` and code `0`.

We can see an example of this in the following screenshot:

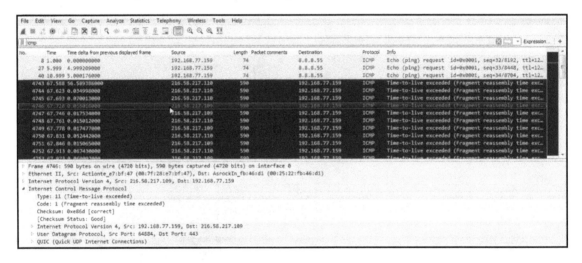

The black packets are highlighted by default in Wireshark. All of them have their TTL exceeded. So there are TTLs that, when they were out on the internet, ran out. They got stuck in a loop or something like that and the router that discarded it sent back a message to me letting me know that my packets have been discarded—they did not make it to their destination.

Now, these packets are a little bit different in that they're fragments. We saw that the other one was a code 0 and the packet shown in preceding screenshot is a code 1, but it's the same idea. It's sent back a type 11 to us to let us know that the TTL was exceeded for that packet. This is actually how traceroute works. It uses ICMP packets with different TTLs in order to figure out what routers are in the path from a source to a destination.

Using traceroute

We'll start a new capture and then traceroute. In Windows, it's tracert. We will traceroute to a different device out there on the internet.

We will traceroute out to our trusty Google DNS server; why not? We will then press *Enter,* and we see the path start to show up here:

```
C:\Windows\system32\cmd.exe                                    ─ ▢ ㎩

C:\Users\sayalit>tracert 8.8.8.8

Tracing route to google-public-dns-a.google.com [8.8.8.8]
over a maximum of 30 hops:

  1      1 ms       1 ms       1 ms   192.168.4.3
  2      *         20 ms      21 ms   arenafirewall.packtpub.net [192.168.4.1]
  3     15 ms      15 ms      15 ms   123.252.235.121
  4      *         22 ms       *      static-2.79.156.182-tataidc.co.in [182.156.79.2]
  5      *         25 ms      21 ms   10.117.225.82
  6      4 ms       4 ms       3 ms   10.117.137.146
  7     23 ms      26 ms      25 ms   14.141.63.225.static-mumbai.vsnl.net.in [14.141.
63.225]
  8      *          *          *      Request timed out.
  9     13 ms       9 ms       8 ms   115.113.165.98.static-mumbai.vsnl.net.in [115.11
3.165.98]
 10     14 ms      15 ms      26 ms   108.170.248.209
 11     16 ms      13 ms      10 ms   209.85.253.211
 12     25 ms      29 ms      29 ms   google-public-dns-a.google.com [8.8.8.8]

Trace complete.

C:\Users\sayalit>
```

You can see that it has received a number of responses as it went through each different router. Every once in a while you'll notice that it gets some timed-out responses. Depending on the router or firewalls that it goes through, they may not send back information to us. This is why sometimes you'll get gaps in your path, but you can see that it picks up again as it finds another router that responds to us at least. Hence, we're still going through the internet, bouncing through routers, and eventually getting to our destination. If you've noticed, each one of these hops was done three times. So, it gives us an average of the response time for that specific router over an average of three attempts—it goes from router to router.

If we take a look at Wireshark, we should see three requests for each router. What it does in my system is, when it creates a traceroute, it starts out by sending a packet to my destination with a TTL of 1, not 128 or 60, or anything like that:

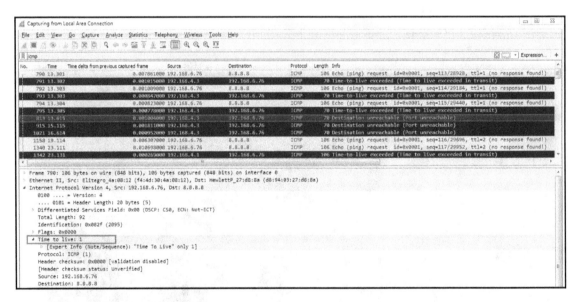

So it'll go to our router that will get it and say, "I can't do anything with this; I need to discard it". The TTLs will go to 0. Then our router will send an ICMP back to me, letting us know that my TTL has been exceeded. What we can see is that's exactly what happens. We sent out a ping request, an echo request with a TTL of 1. Our router responded to say "oh, sorry; I couldn't do it". Then, I do it again for that second attempt. We get a response that says "no, I can't do it". This is done a third time, as well. Then, it goes and creates an echo request with a TTL of 2:

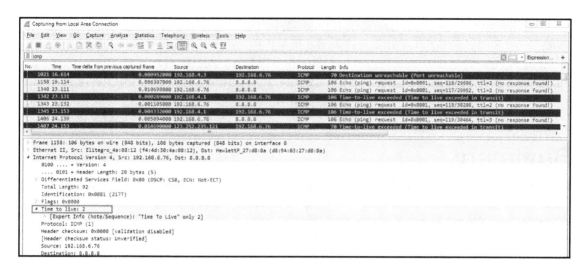

It gets through our router because my router sees that it's a 2; it decrements it to 1 and forwards it; that's completely valid. Then it goes to the next router up the chain, on Verizon, we saw. That router sees it; it sees that it's a TTL of 1; it has to discard it and send back a response saying "no, sorry; I couldn't do it; your TTL has exceeded". I do it again and it responds. Now, we create a TTL of 3:

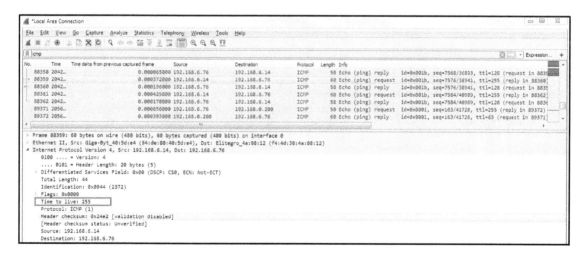

It keeps doing that for TTL 4 and so on. That's how traceroute is able to map out the routers from point A to point B—because it uses the ICMP TTL exceeded errors in order to figure out that device. This is because it knows that the ICMP packet is generated by the router as an error code back to us. Luckily, we can use that IP information to map out the hops that I go through in order to get to that destination. Very clever.

Summary

In this chapter, you learned how DNS functions and saw some DNS analysis. We saw how ARP works and went on to resolve the MAC addresses. We covered IPv4 and IPv6 headers, and saw how to take a look at some of the details within the headers, including ICMP. We also saw why ICMP is useful to us as network engineers. We covered some information on how traceroute works, and the headers in it, as well.

In Chapter 7, *Network Protocol Analysis*, we'll start diving more into analysis and taking a look at some additional protocols and applications.

Network Protocol Analysis **7**

In this chapter, we'll take a look at the following topics:

- UDP analysis
- TCP analysis
- Graphing I/O rates and TCP trends

UDP analysis

We'll take a look at how UDP works, what it is, and what's in the UDP header. The UDP protocol is a connectionless protocol and it's very lightweight—a very small header.

If you'd like to learn more about the UDP protocol, take a look at `https://www.ietf.org/rfc/rfc768.txt`:

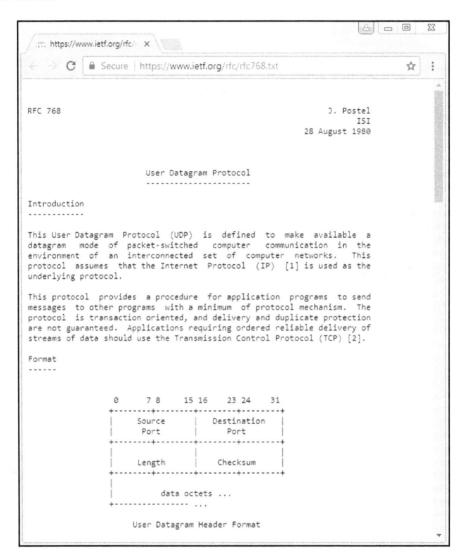

This is the original specification. It's been updated since August 28, 1980, if you look through all of the RFCs, but the original specification is `768`. If you'd like to learn about all the details of UDP, which are relatively short, you can do so through the file shown in the preceding screenshot.

Let's take a look at UDP in Wireshark:

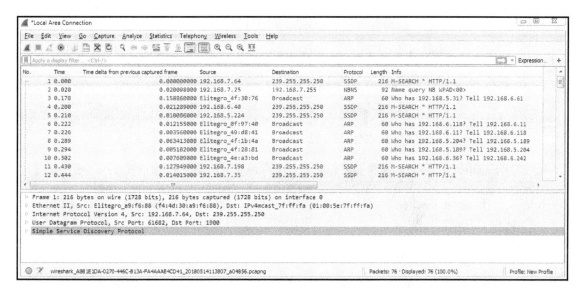

We have a capture of just a few seconds of data and a whole mixture of applications and protocols. What we can do is simply filter based on udp. If you press *Enter*, now it only shows UDP packets:

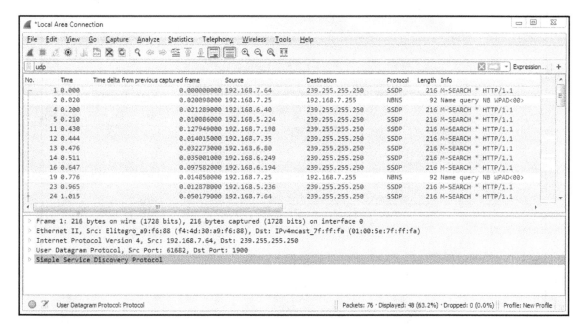

And you can see that we have some additional protocols listed, which include applications that use UDP for their transfer, such as `SSDP`. What we can do in the packet details is take a look at the UDP section:

```
◢ User Datagram Protocol, Src Port: 61682, Dst Port: 1900
        Source Port: 61682
        Destination Port: 1900
        Length: 182
        Checksum: 0x4b86 [unverified]
        [Checksum Status: Unverified]
        [Stream index: 0]
```

In the UDP header, there are very few fields. UDP always has 8 bytes in its header, and there are only four fields. We have a **Source Port**; a **Destination Port**; the **Length**, which is the total length of the packet, including header and data; and a **Checksum**, which validates the header information. But it does not encompass all of the data like you would expect, with the FCS in a frame, at the end of a frame. You can see that we have an `unverified` checksum. By default, this option is not enabled in Wireshark.

Now, go to **Edit** | **Preferences...** | **Protocols** | **UDP** and turn the **Validate the UDP checksum if possible** option on:

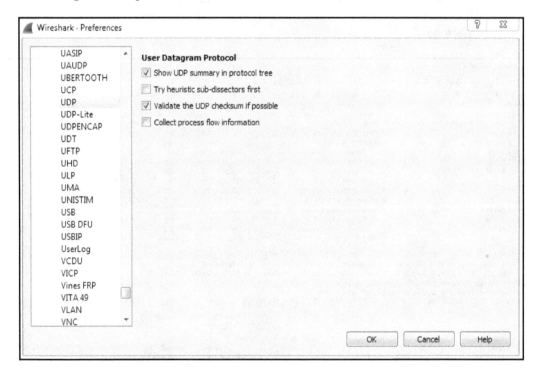

You can see that now the checksum says `correct`:

```
⊿ Checksum: 0x9443 [correct]
      [Calculated Checksum: 0x9443]
   [Checksum Status: Good]
   [Stream index: 10]
```

So, if there were any problems in the header and it was manipulated in transfer, we'll be able to see that here and it'll be marked. Then, if you expand **Checksum**, it'll tell you what is the checksum information that it was calculating.

There are usually very few problems that you'll have with a UDP transfer. Either they work or they don't. They do not guarantee any connectivity, and the applications will perform any sort of retransmission if necessary, built into the application. It's not handled within the stack like it is with TCP. Because it has a very small header and very few fields, there are very few options to be turned on and off. There's not much here; it's meant to be very simple and lightweight, which is great for voiceover IP or streaming video; anything like this, which is very time-sensitive. It sends the data on its way and hopes that it gets there. Great if it does; if it doesn't, then: oh well, you miss a packet or two.

One thing you can do if you're not sure whether a packet is UDP or not when looking through the packet list up top is create a column based on UDP. So we right-click on **User Datagram Protocol, Src Port: 40097, Dst Port: 1900** and select **Apply as Column**:

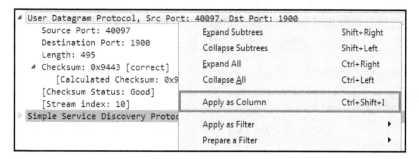

Now, we have a new column that says **User Datagram Protocol**; it is a UDP packet. If we remove our udp filter, we can see that we now have a tick mark and blank listed throughout our capture:

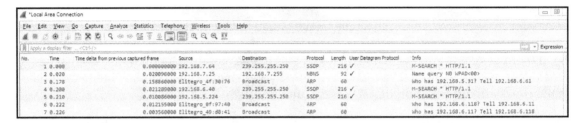

That's a nice way of easily seeing what is a UDP packet and what isn't, depending on whether or not you have different coloring rules or something like that that's in a large packet capture.

One of the few problems you may see with UDP are destination unreachable responses (these are ICMP packets, if you remember) following UDP connection attempts. If you have a UDP connection attempt and you continuously receive an ICMP destination unreachable in the next packet and also later on, that's an indication that you might have some sort of connectivity issue that you need to investigate. That's really the only sort of response you'll get because UDP does not send responses. The device itself may send a response telling you that a network is unavailable or something like that and hence the destination is unreachable. But, otherwise, UDP itself will not tell you anything. This is why there are very few things that you will see in a packet capture regarding UDP issues because there's nothing built into UDP to tell you that there are issues.

TCP analysis I

In this section, we'll take a look at how TCP works, what's in the TCP header, and some of the flags and options.

If you'd like to learn more about TCP, you can look at the RFC that's available from the IETF at https://tools.ietf.org/html/rfc793:

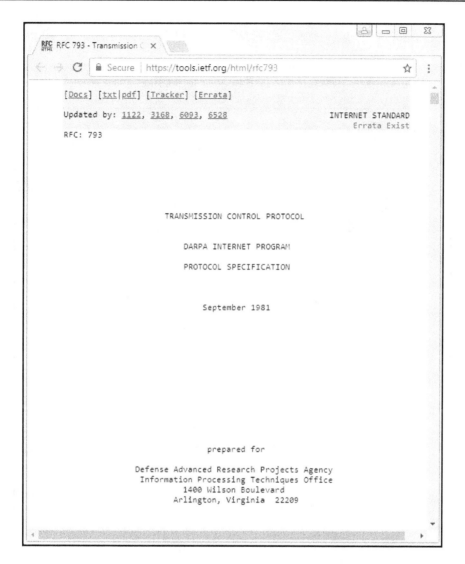

You're looking for **RFC: 793** for TCP, which is the original specification for TCP.

In the preceding screenshot, you can see different sections within IETF, which provide a little bit of interactivity. You can click on the different RFCs that have updated the TCP specifications; if you scroll down, it also provides you a nice little table of contents. The RFC shows a little diagram of what the TCP header looks like:

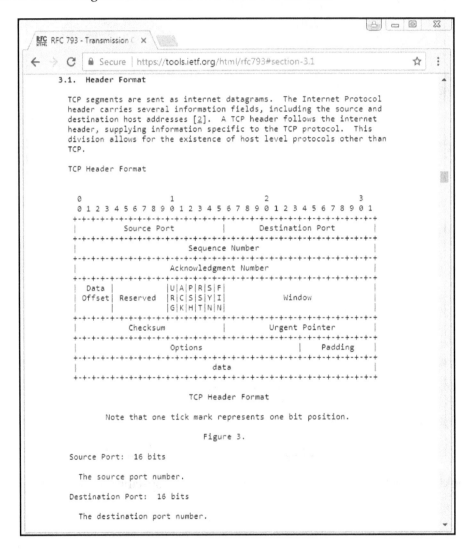

We have **Source Port**; **Destination Port**; **Sequence Number**; **Acknowledgment Number**; **Data Offset**; some **Reserved** bits; **Window** size; header **Checksum**; an **Urgent Pointer**; and **Options**, which is an expandable section. We have some **Padding** and then the actual **data**.

Go into Wireshark and let's go to a TCP packet. We can see we have some TLS traffic that happens to be running over TCP. We can right-click on **Transmission Control Protocol** and select **Apply as column**:

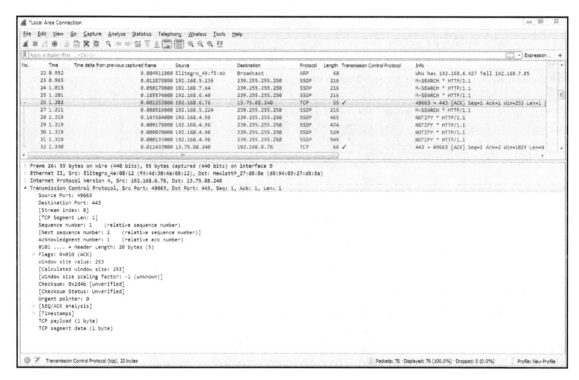

In the preceding screenshot, you can see the tick mark, which means it is indeed using TCP. If we expand **Transmission Control Protocol**, you'll see the fields that we were just looking at in the RFC.

We have **Source Port** and **Destination Port**, and you can see that the system in my computer was accessing this web resource on port 443 and accessing other fields, which you've noticed in other packets that you've looked at, most likely. Anything with a square bracket is created by Wireshark; they're not actually fields within the header, so we can skip them. We have **Sequence number** and **Acknowledgment number**. These are relative numbers, if you noticed. We mentioned this a little while back that they started with 1 in this example. The actual raw number is not 1. Wireshark shows it as 1, and as a relative number to make it easier for you to look at. Otherwise, it's a very long number that is harder for humans to look at and diagnose.

Next, we have **Header Length**. It tells us how big the header is because the header can change its size in TCP, unlike UDP, so we have to tell it how long it is. We have some **Flags**, as shown in the following screenshot:

```
⊿ Flags: 0x010 (ACK)
     000. .... .... = Reserved: Not set
     ...0 .... .... = Nonce: Not set
     .... 0... .... = Congestion Window Reduced (CWR): Not set
     .... .0.. .... = ECN-Echo: Not set
     .... ..0. .... = Urgent: Not set
     .... ...1 .... = Acknowledgment: Set
     .... .... 0... = Push: Not set
     .... .... .0.. = Reset: Not set
     .... .... ..0. = Syn: Not set
     .... .... ...0 = Fin: Not set
     [TCP Flags: ·······A····]
```

In the preceding screenshot, you can see we have some congestion information; the **Urgent** bit; **Acknowledgment** and **Push**; and **Reset**, **Syn**, and **Fin**. A lot of this stuff will look familiar, such as **SYN**, **Acknowledgment**, and **FIN** with creating a connection and finalizing a connection. We also have **Window size value**, which tells us how large of a chunk of data we can transmit before having to do an acknowledgment:

```
Window size value: 253
[Calculated window size: 253]
[Window size scaling factor: -1 (unknown)]
Checksum: 0x2d4b [unverified]
[Checksum Status: Unverified]
Urgent pointer: 0
▷ [SEQ/ACK analysis]
```

We have a **Checksum**, which again is unverified, and an **Urgent pointer**.

If we want, we can go to **Edit** | **Preferences...** | **Protocols** | **TCP** and enable the **Validate the TCP checksum if possible** if you would like:

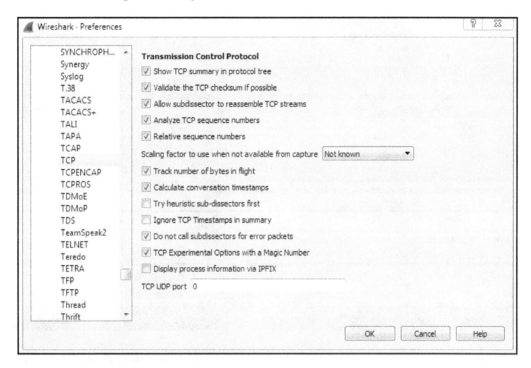

We can see that we now have a `correct` checksum.

Let's go ahead and create a new capture, and we'll take a look at a handshake creation and teardown of a connection.

Let's start a new capture, and we will generate some traffic. Furthermore, open up a new web page and re-download all the information for that RFC that I showed you from the IETF website. We'll scroll down and look for the beginning of our connection:

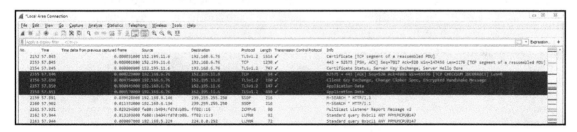

It looks like the website that's being transmitted. It was over HTTPS, so seeing the TLS traffic is what we can expect. We'll right-click on the TLS traffic and go to **Follow** | **TCP Stream**:

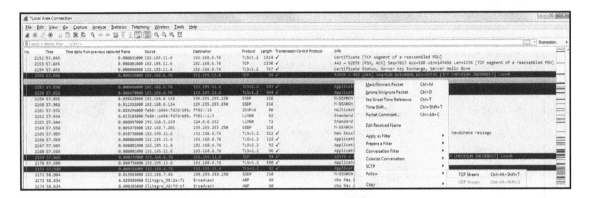

Next we'll just go back to our main packet list:

You can see that the filter for `tcp.stream eq 27` is listed in the filter option. Now this is our connection to the website for IETF. As mentioned earlier, TCP is connection-oriented. It guarantees delivery of packets. So if you miss a packet, the acknowledgment system builds in and retransmits it. So what happens is that we send a piece of data from the server to the requesting device. We send an acknowledgment that we received it. We also send another packet followed by an acknowledgment indicating that we've received it.

Sometimes, depending on the options and the functionality of the TCP stack in your network card in your drivers, you may allow for some enhanced features that have come out over the years, such as selective acknowledgments and some enhancements to allow you to be more efficient with the way of using this acknowledgement system so that you don't have as much overhead. The creation of a TCP handshake we can see is a `SYN`, which says "let's synchronize":

```
◢ Flags: 0x002 (SYN)
      000. .... .... = Reserved: Not set
      ...0 .... .... = Nonce: Not set
      .... 0... .... = Congestion Window Reduced (CWR): Not set
      .... .0.. .... = ECN-Echo: Not set
      .... ..0. .... = Urgent: Not set
      .... ...0 .... = Acknowledgment: Not set
      .... .... 0... = Push: Not set
      .... .... .0.. = Reset: Not set
   ◢ .... .... ..1. = Syn: Set
      ▷ [Expert Info (Chat/Sequence): Connection establish request (SYN): server port 443]
      .... .... ...0 = Fin: Not set
      [TCP Flags: ··········S·]
   Window size value: 8192
   [Calculated window size: 8192]
```

Let's create our connection. We have a `Syn` in the **Flags** that is `Set`. That's my system requesting the IETF website create a TCP connection. They respond with a `Syn` and an `Acknowledgment`; then we, in turn, send back an acknowledgment, as well. So we have `SYN`; `SYN, ACK`; and `ACK`—the three-way handshake with TCP. After that occurs, we then begin communication to retrieve HTTP traffic, which happens to be laid underneath TLS, in this case. We can see the creation of the TLS connection and some data being transmitted, such as key exchanges and things like that:

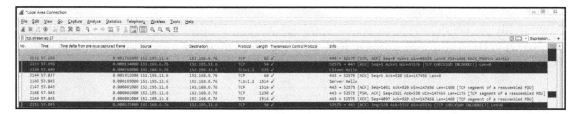

Then, at the end of the communication, while sending and receiving whatever data we wanted such as the website being retrieved, our system will then finalize that and with an implicit method say: "let's terminate this; there's no more data to be sent or received". We'll end this TCP connection with a finalized statement.

And so what we do is, in the **Flags**, we have `FIN` and `FIN, ACK`. Sometimes at the end of the list, you'll also see a **Reset**. So the final method using `FIN` is implicit, saying this connection should be terminated; there's nothing else to be transmitted. But it's not explicitly told to remove the connection, so the system on either end may end up leaving this connection live—it's up to them.

If you see a **Reset** at the end of a connection after the FINs, then that's an explicit way of one of the devices saying: "yes, terminate this packet". This is an explicit way of saying: "kill the connection; reset it; nullify it".

If you see a **Reset** before the FINs, then that may be an indicator of a problem in the connection, where the server or the requesting device solves a problem and resets the connection to try again.

We also have options, which are an extension in the header to allow us to expand upon the abilities of TCP. What we can do is filter on options because, if you notice looking through the header, we don't see anything that says options. If we want to find anything that has options, we need to filter on that.

In the display filter, type tcp.options:

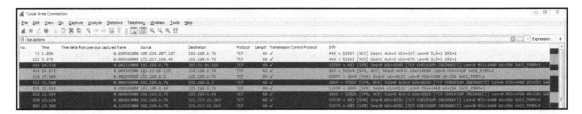

Now we're showing all the traffic that has options enabled in the header. If we scroll down in the packet details area, we have some information about the **Options**. We can skip the **No-Operation** stuff. We want to look at things that are more interesting to us, such as **TCP Option - SACK permitted** and **TCP Option - Window scale**:

```
▲ TCP Option - Window scale: 8 (multiply by 256)
      Kind: Window Scale (3)
      Length: 3
      Shift count: 8
      [Multiplier: 256]
▷ TCP Option - No-Operation (NOP)
▷ TCP Option - No-Operation (NOP)
▲ TCP Option - SACK permitted
      Kind: SACK Permitted (4)
      Length: 2
```

For example, the following screenshot shows a packet that has TCP SACK:

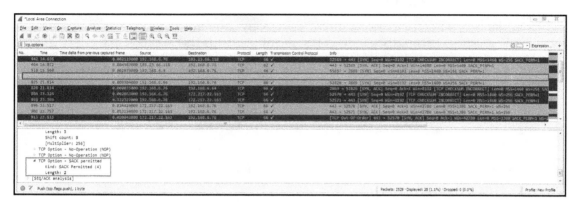

Selective Acknowledgment (**SACK**) allows you to acknowledge traffic while still requesting for missing traffic, without having the entire block of packets resent to you. So if you're having, let's say, five packets at a time being sent to you before you send an acknowledgment and one of these is missed, without Selective Acknowledgment all five will need to be retransmitted to you. So, the further apart the acknowledgments are between the number of packets that are sent to you, the more efficient it is because you're not wasting all this bandwidth with overhead acknowledgment packets. But if you run into a problem, then more data has to be retransmitted. Selective Acknowledgment is a way of getting the best of both worlds so that you can have large blocks of packets being sent to you without having to acknowledge them every single time but, if you miss one, you can then request just that one to be resent to you; you can still acknowledge all the other packets. If you see TCP Selective Acknowledgments, it means that both the devices have allowed for Selective Acknowledgments and they've agreed upon that. Both sides will have to allow for that feature set to be enabled in order to utilize it.

We also have window scale. Window scale, as shown in the earlier screenshot, allows us to go beyond the initial window size. The initial window size was at a maximum of 65535 bytes, which isn't a lot anymore, so we want more than that. We can do that using window scale. Window scale allows us to multiply the window size by a factor. So you could say 65535 times whatever the window scale is, for example. Then, you could get a very, very large window size in order to most efficiently use your bandwidth.

TCP analysis II

In this section, we will take a look at filtering on many different TCP header fields, and what kind of issues we could see based on some of the fields that we look at.

So, what we can see in the following screenshot is a packet capture of the websites that were opened:

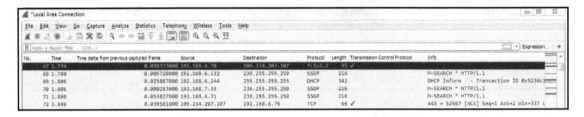

You can see we still have the TCP column enabled, so we can go down and find a TCP packet, and we'll see the field information again:

```
⊿ Transmission Control Protocol, Src Port: 443, Dst Port: 52567, Seq: 1, Ack: 2, Len:
      Source Port: 443
      Destination Port: 52567
      [Stream index: 0]
      [TCP Segment Len: 0]
      Sequence number: 1      (relative sequence number)
      [Next sequence number: 1      (relative sequence number)]
      Acknowledgment number: 2      (relative ack number)
      1000 .... = Header Length: 32 bytes (8)
   ▷ Flags: 0x010 (ACK)
      Window size value: 337
      [Calculated window size: 337]
      [Window size scaling factor: -1 (unknown)]
      Checksum: 0xc64b [correct]
      [Checksum Status: Good]
      [Calculated Checksum: 0xc64b]
      Urgent pointer: 0
   ▷ Options: (12 bytes), No-Operation (NOP), No-Operation (NOP), SACK
   ▷ [SEQ/ACK analysis]
   ▷ [Timestamps]
```

As with many of the other protocols that we've looked at, we can right-click on any of the fields and apply them as a filter. What we may want to do is expand the **Flags**, and look for anything that has the urgent bit set:

```
⊿ Flags: 0x010 (ACK)
      000. .... .... = Reserved: Not set
      ...0 .... .... = Nonce: Not set
      .... 0... .... = Congestion Window Reduced (CWR): Not set
      .... .0.. .... = ECN-Echo: Not set
      .... ..0. .... = Urgent: Not set
      .... ...1 .... = Acknowledgment: Set
      .... .... 0... = Push: Not set
      .... .... .0.. = Reset: Not set
      .... .... ..0. = Syn: Not set
      .... .... ...0 = Fin: Not set
      [TCP Flags: ·······A····]
```

The urgent bit is not often used. The only one that we can think of is using Telnet, and what it does is it prioritizes the packets, basically. So what we can do is right-click on **Urgent** bit, and we'll go to **Apply a Filter | Selected**:

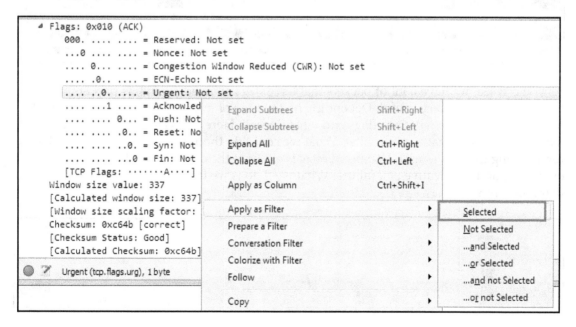

And we'll see that the filter's created, but it's based on 0. So we'll change it to 1:

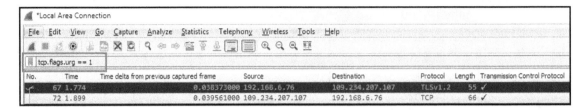

Now, we will see whether there are any packets that have the urgent bits set. We see that there are none, which is good. Another thing we can do is delete the filter back to `tcp.flags`, and if we press *Enter* on that, it now filters based on every single packet that has flags enabled:

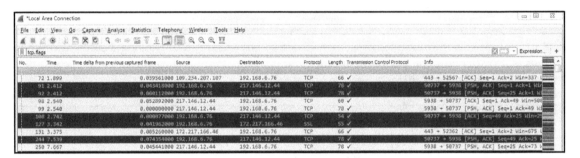

As you can see, we have a lot of packets that have flags enabled. So in this packet that capture may not be very useful. Depending on the type of traffic that you are seeing, that may be helpful to see what has flags and what doesn't. Here, it has too many packets that have flags to make that a useful filter. What we could do though is customize that and look for anything that has a reset. Usually, a reset is either at the end of a good connection, as explained, or it's indicative of a failure. What we'll do is we'll change our display filter so that it is `tcp.flags.reset == 1`:

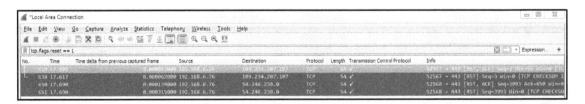

You can see we have quite a few packets that have resets. We can take a look at one of these, and look at the traffic surrounding the reset to see whether that's indicative of a problem. So, let's choose one packet and right-click on it and go to **Follow | TCP Stream**:

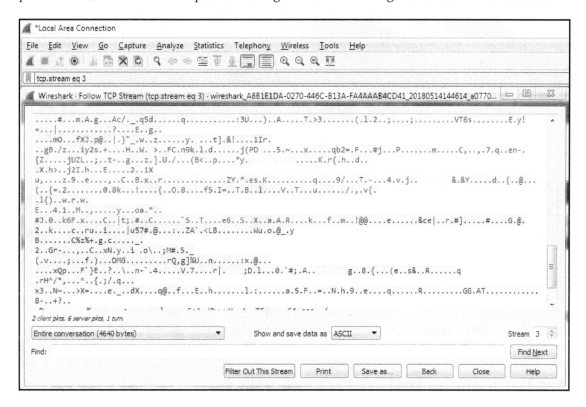

Close this, and we see it looks like some sort of a certificate transfer:

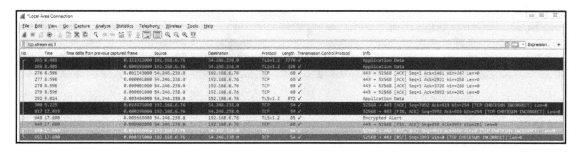

We have a three-way handshake; we have a client and server hello—it looks like it was a TLS negotiation. Then there are some packets of application data. Furthermore, the system sent a FIN, ACK, so we finalized the connection. That was an implicit termination, which was then followed up with an explicit termination, the reset. So this is actually fine. The system said: "I'm done with this connection, and then we'll reset and totally terminate this connection". So that's actually fine, but you can see how that's useful to be able to pull out everything that is a reset.

What we could also do is take a look at window size, and for that we will enter tcp.window_size < 50 and press *Enter*:

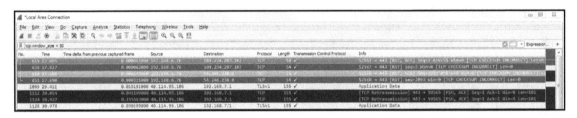

We see that we have quite a number of packets that have a window size of less than 50. Usually, a small window size is an indication of a problem. On a reset packet, it's not necessarily a problem because we're terminating a connection anyway, but if you see window sizes that are of a small value on standard data transfer packets, that's a problem. That's some sort of buffer problem in your linear devices in the network stack on, on one of the two. Additionally, Wireshark has some analysis filters that we can use, and we'll show one that is window size related. For that we will enter tcp.analysis.zero_window. If we press *Enter* on that, you'll see the following screenshot:

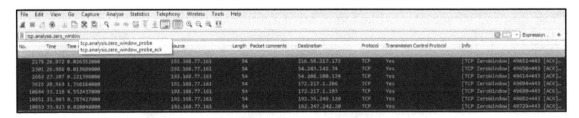

This is actually all of the packets that have a window size of 0, which is a problem. So these are potential issues. These are acknowledgments from the system to a variety of different servers out there, with the system declaring that it has no buffer space available. It's received buffer's full and it sends a zero window response back to the servers out there, saying: "please slow down". So this is a potential issue on the system.

Another interesting one that we could do is take a look at the header length. As you know, the TCP header can fluctuate in size, let's right-click on **Header Length** and go to **Prepare a Filter** | **Selected**:

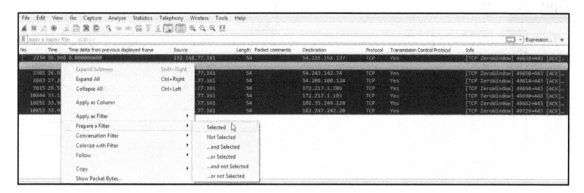

We didn't use **Apply as Filter** here; that way, it doesn't apply it right away.

In the filter tab, we will enter `tcp.hdr_len > 20`, and here we go:

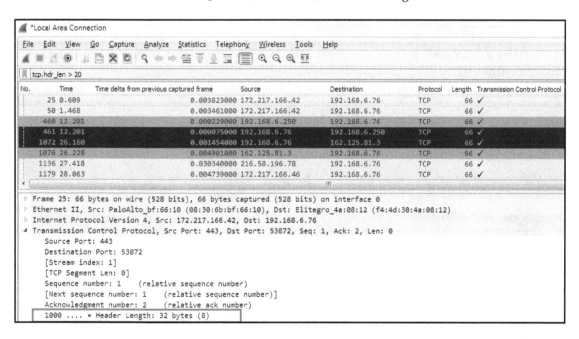

You can see that the first header length that pops up has `32 bytes`. So, here's a good example of finding packets that have **Options** in them.

If we scroll down, we can see that the **Header Length** is larger than the standard 20 because it has **Options** built into it somewhere. Here, you can see this packet has Selective Acknowledgments and window scaling. You could also do a similar thing by filtering on **Options**, so right-click on **Options** and go to **Prepare a Filter | Selected**:

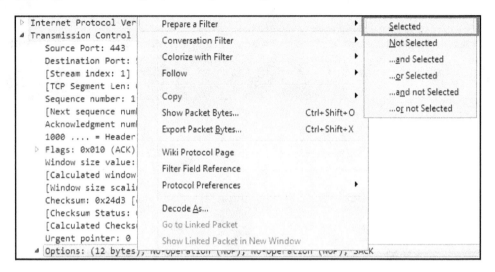

We prepare a filter and enter `tcp.options`. So here's every packet with options that are enabled within them:

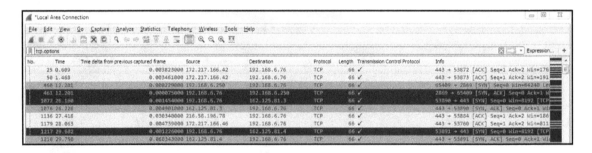

Graph I/O rates and TCP trends

In this section, we'll take a look at using graphs to help visualize packets and trends, especially in TCP communications. Here we have a packet capture of a file transfer that has gone horribly wrong:

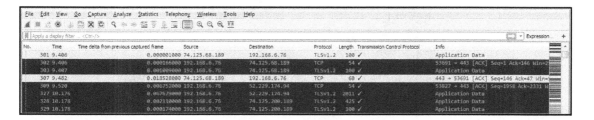

As you can see, we have all sorts of black bars coming up, which is a big telltale sign in Wireshark that there's something amiss. You can also see there's a striping pattern that comes about. That's a big telltale sign that there are a bunch of retransmissions. So, what we'll do is use this as our basis for graphing and being able to pick out some issues.

Throughput

One of the first things we'll do is go up to **Statistics | TCP Stream Graphs | Throughput**:

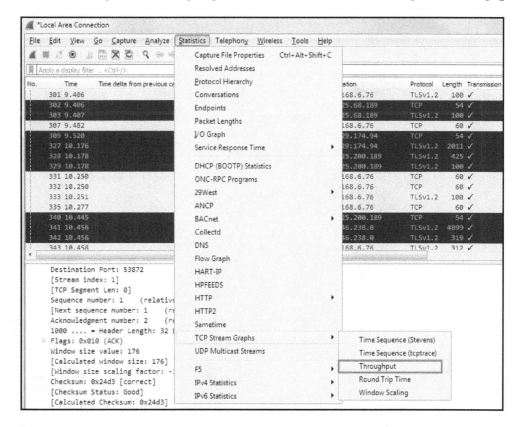

When we click on **Throughput**, we will see that a graph comes up:

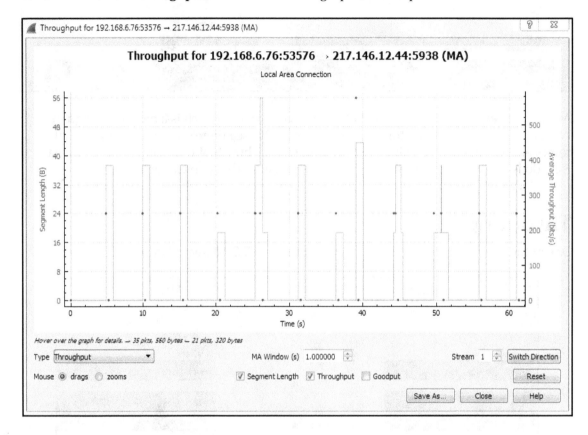

Whether it's graphing I/O rates going into the section, going for TCP stream graphs, or anywhere like that, all the graphs are unidirectional. Depending on what packet we have selected, it will show us the throughput for that or the I/O rate, or whatever it might be that we are graphing—it'll be for that one unidirectional transfer. As you can see we'll go from `192.168.6.76` to the public address as we're trying to pull HTTP. If you don't see what you're looking for in the graph, click on the **Switch Direction** button:

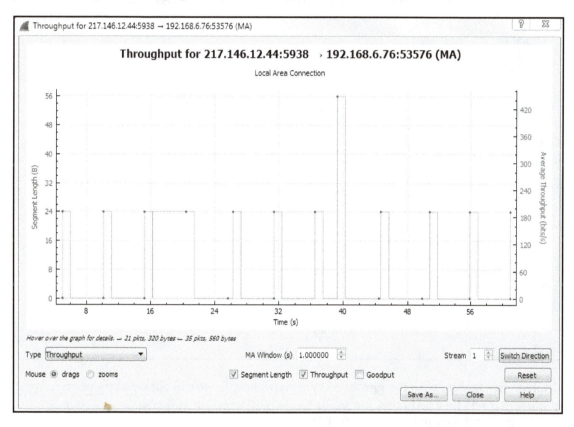

This'll provide us with the opposite direction. We can see here that we're getting the throughput from public address with port `5938`, the server, and that it's going to our private address that was the web browser. We can see that we have the throughput graphed. We have the **Segment Length** and we have the throughput itself in bits per second. What you can do, in the latest version of Wireshark, is you can drag the graph around in order to be able to view it more effectively or you can use your scroll wheel to zoom in. Alternatively, you can use the plus or minus signs on the right-hand side of your keyboard or in the keypad area. I like to use the scroll wheel just because it's handy and almost every mouse has it.

Also note that you can change the **Stream** that you're following. If there were multiple streams that we wanted to look at, we could change the stream number that we're in. Here, we'll leave it as 1:

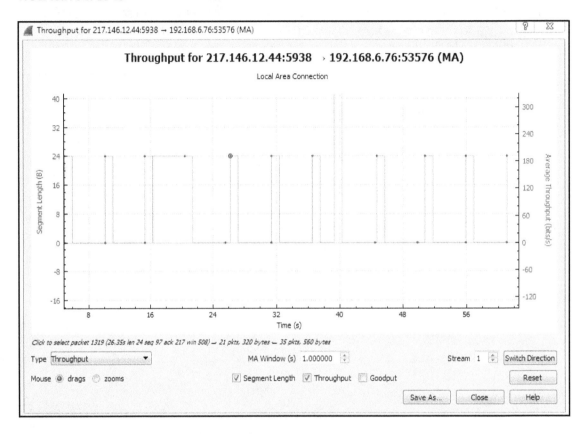

Another nice feature with the graphs in Wireshark here is we can click on any point in the graph and it will take us to that packet. Let's click between the little gap seen in our throughput, and we can see that Wireshark takes us to that section of packets.

If you noticed, when we went to **Statistics | TCP Stream Graphs**, there were several options to choose from:

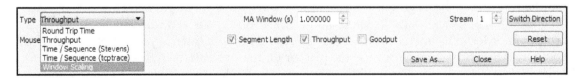

You don't have to close this window and go back to the **Statistics** window and open it again; you can switch between them easily by simply selecting whatever type you want.

Let's go to **Time / Sequence (Stevens)**:

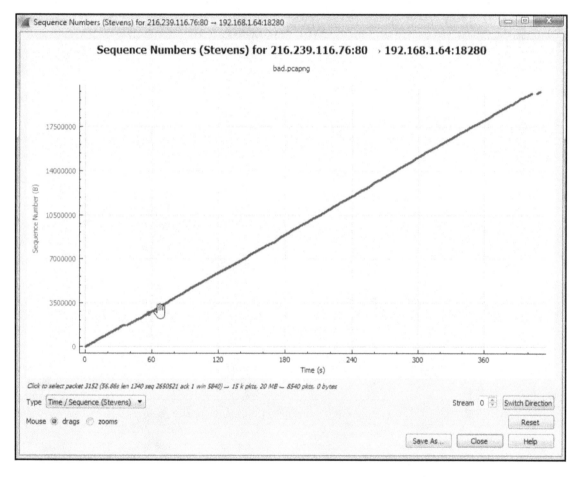

You see here that we have a nice diagonal line in our time sequence. What that's showing us is the sequence number continually incrementing from packet to packet. Sometimes, you can view issues in your packet capture with sequence numbers by seeing a drop in sequence numbers or maybe a flat line, where it does not increment upwards in the y axis and it just continues on the x axis. If you see anything other than a diagonal line, that's indicative of issues in your packet capture. So what we see here looks relatively good. We can close it.

I/O graph

Now, let's go to **Statistics** | **I/O Graph**:

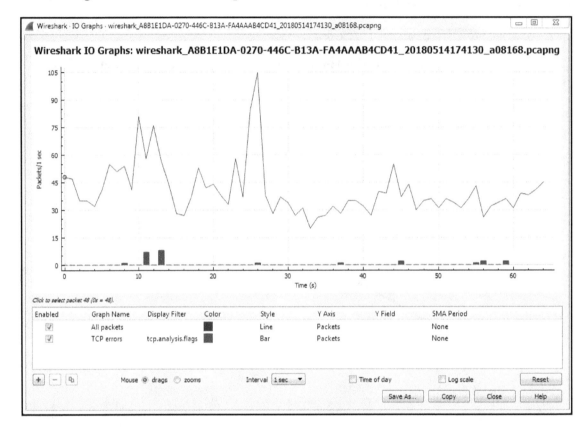

The I/O graph is a very powerful feature in Wireshark, and here it's showing us all the packets per second, and all of this is customizable. You can change the **Interval** and you can also change whether it's a linear or logarithmic scale:

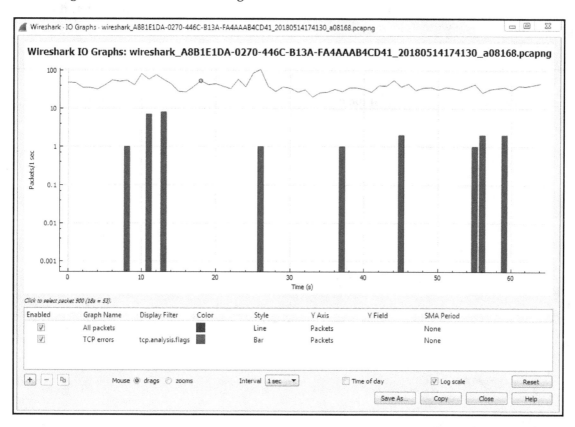

Depending on the needs of the packet capture that you're looking at, maybe there are very drastic differences between some different streams that you're wanting to graph, or maybe some differences between the filters that we want to apply because you can layer filters in this section. You may want to use logarithmic for that; it all depends on what you need the packet capture to look like. By default, when you open it up, Wireshark will show you the number of packets per second, and then every tick is one second. What we have are a number of additional layers that we can customize. You can see that we have a **Name** column, where you can name the layer. We have a **Display filter**, which includes **Color**, where you can change the color and the **Style** type, whether it's a line graph or a bar graph. You can change what the y axis represents. You can turn on **Smoothing**; you can do a number of different adjustments here to your I/O graph to make it as customized as you need.

One powerful thing to look for in a packet capture that you know has a problematic transfer are TCP analysis flags. We'll do that for the second layer. For that we'll create a display filter for TCP analysis flags, and layer that on to our I/O graph. So, what we'll do is double-click on the second **Display filter**, which allows us to enter a display filter. It will work just as if you were adjusting the display filter in the main packet list area. We will add `tcp.analysis.flags`:

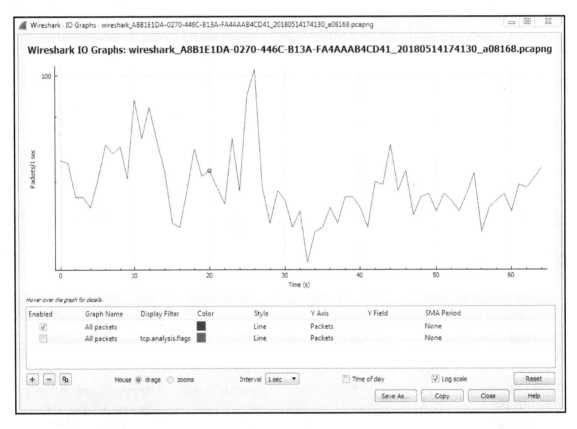

When we apply this, it will create a second layer on the graph with a red line graph of any packet that has an analysis done to the flags. Wireshark looked into the flags and saw that something occurred and it provided some sort of insight into it. Just like we saw with the expert information in the lower left, this display filter uses the same functionality as the expert information.

What we don't want to include, though, are window updates because window updates are good. As long as it's not a zero window situation where the window size is 0 and our buffers are full, we don't want to include all the packets that have window updates.

Next, we will negate window updates from this. We will add `tcp.analysis.flags &&`. If we use an exclamation mark, that will negate whatever we are about to include. It will be `tcp.analysis.flags && !tcp.analysis.window_update`.

When you're done, simply click away and it will apply the filter.

Now you can see that we have the check marks on the left that will enable or disable the following layers:

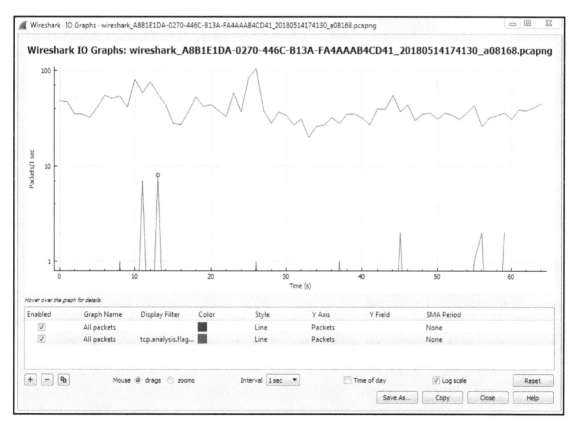

So, if we disable the first packet option, that gets rid of our packets per second. If we enable the second one, that will show us all of our TCP analysis flag issues. You will see that there are a number of problems.

What we can do is, make this easier to view and really make it stand out, especially if there's a lot of data going on. If we have a lot of additional layers, sometimes you may want to make certain layers stand out. What we can do is change the **Style** of it to a **Bar**:

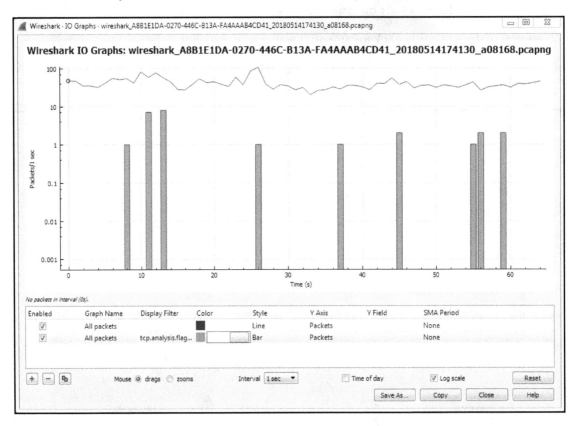

Once we change it to a **Bar**, you'll see that it's a bit thicker and stands out more.

And if you zoom in, you'll actually be able to see the pink there, and it really sticks out like a sore thumb. Then, we can click anywhere along this line and it will take us to the problem:

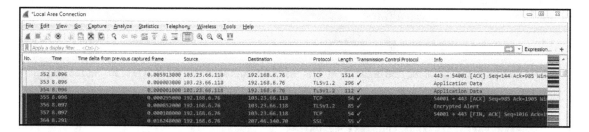

Now, we can see all of our TCP analysis issues. So what we can do is, click on that packet and validate this, as well. If we click on one of these black packets, expand TCP, and then look at the analysis section, it tells us that there's a duplicate. This is a duplicate to a previous acknowledgment, which is obviously a problem that's indicative of an issue:

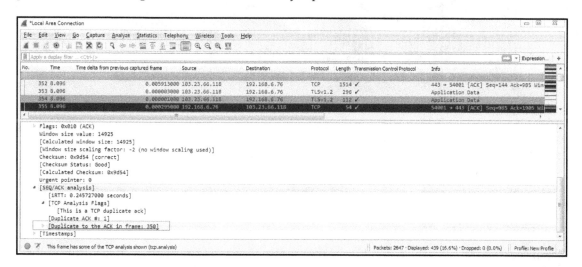

Also, note that these different layers are on top of each other, so the top layer is the one that's closest to the foreground and then the one towards the bottom is towards the background. If you have a lot of data on a foreground layer, it may overlap and overwrite visually what's going on in a background layer, so be aware of that. For example, consider changing the style of this to a **Bar** chart :

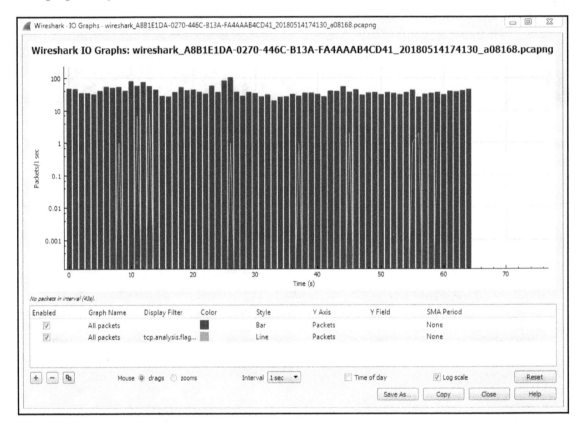

You'll see that it makes the second layer that we created with the analysis flags useless because it is now dominating the entire view. So be careful with the way you change and use your different layers.

Summary

In this chapter, you learned about the UDP protocol and its connectionless abilities—the very small header that it has, all the different fields in the TCP header, and the options that are available in it.

We also looked at TCP, the connection-oriented protocol, and the guaranteed transmission of certain data. We looked at the header and learned about the fact that it has different flags and different options to expand upon its capabilities. We also learned about the graphing functionality in Wireshark, and the fact that you can add multiple layers and change the way the lines and bar charts work, in order to get the most useful graph out of it.

In Chapter 8, *Application Protocol Analysis I*, we will take a look at HTTP and FTP, and some of the common applications that you'll run across on a day-to-day basis.

8
Application Protocol Analysis I

In this chapter, we will cover the following topics:

- DHCP analysis
- HTTP analysis I
- HTTP analysis II
- FTP analysis

DHCP analysis

In this section, we'll take a look at how DHCP works, some of the fields that are within the DHCP protocol, watch a client retrieve an IP address, and also take a look at what happens with DHCP when a client requests an address and receives responses.

Let's start a packet capture. What we'll do now is release the address on my computer and then renew it.

Type `ipconfig /release` on a Windows computer on Command Prompt to release our address, then if we type `ipconfig /renew`, it will get us a new address.

Now, if we type `ipconfig /all`, we should be able to see that our address is assigned. We'll stop our capture now. We'll want to only pick out the DHCP traffic. So you would assume you could go up to the display filter and type `dhcp`, just like we've done for the other protocols, and then press *Enter* and it works. But we can see that there's a red bar up there, which indicates that `dhcp` is not valid:

No.	Time	Time delta from previous captured frame	Source	Destination	Protocol	Length	Transmission Control Protocol	Info
1	0.000	0.000000000	192.168.6.108	239.255.255.250	SSDP	175		M-SEARCH * HTTP/1.1
2	0.020	0.020683000	192.168.6.72	192.168.7.255	BROWSER	243		Host Announcement PPIUMCPU0357, Workstation,
3	0.064	0.044158000	192.168.5.226	239.255.255.250	SSDP	216		M-SEARCH * HTTP/1.1
4	0.099	0.035141000	192.168.5.207	239.255.255.250	SSDP	216		M-SEARCH * HTTP/1.1

This is because the display filter is actually `bootp`. DHCP is based off of `bootp`. `bootp` was the predecessor protocol to DHCP, so in Wireshark they use the predecessor's protocol filter. Hence, you want to use `bootp`. If we use `bootp`, we'll see that we have our DHCP release and discover:

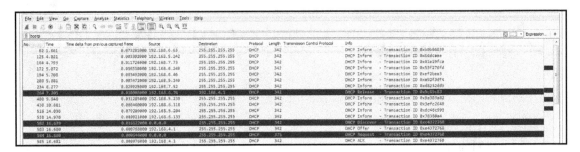

The 264th packet is `Release`. My system already had an IP address, and it wanted to get rid of it. It sent a packet to the DHCP server, which is `4.1`, and it said: "Please get rid of my address." If we look into our DHCP details in the packet details section, you can see that it says **Bootstrap Protocol** (that's where `bootp` comes from) and we scroll down to the bottom and you see options, and if you look at option `53` it says we have a `Release`:

```
▷ Option: (53) DHCP Message Type (Release)
▷ Option: (54) DHCP Server Identifier
▷ Option: (61) Client identifier
▷ Option: (255) End
  Padding: 0000000000000000000000000000000000000000000000000000000...
```

So that's where it's requesting to get rid of its address. The system then at that point has no address, and the local client erases the IP address from its information on the network interface card.

After this, I initiated the DHCP renew command, which told it to go get an address. Now I did this because my system had already been up and had already retrieved a DHCP address on `bootp`. Commonly, a system will retrieve IP information and other configuration options from a DHCP server on boot of the operating system. But since my system had already had an address, I had to get rid of it, and then forcibly tell it to get a new one with that /renew command. When a system requests an address, it initiates with a `Discover` request. And you can see that the discover (which is down at the packet details at option `53` `Discover`) it sent it out to a broadcast to `255.255.255.255`. That's because the client doesn't have an IP address, and you can see that it has `0.0.0.0` as its source.

It doesn't know where it needs to go to talk to the DHCP server. So it sends it out to a broadcast address, hoping that someone will respond to its request for a DHCP server. We can see that the `Discover` packet is asking for a DHCP server. It says: "I'm trying to discover a server." If you have multiple servers on a subnet, you may get a different server offering itself from time to time. This can cause problems sometimes, depending on your network design. If you have a simple network, such as a home network, and you have two DHCP servers, most likely one of them is by accident. You might face that problem when people bring home wireless routers or something like that into a workplace and they don't turn off the DHCP server; it can cause problems like this.

Looking for `Offer` packets sometimes is useful in packet captures because you may not necessarily want to see them; this may be a bad thing. If you happen to see `Offers` from a server address that doesn't make any sense according to your network design, then that's a red flag. You can of course right-click on the **DHCP: Offer (2)** and go to **Prepare a Filter** | **Selected**:

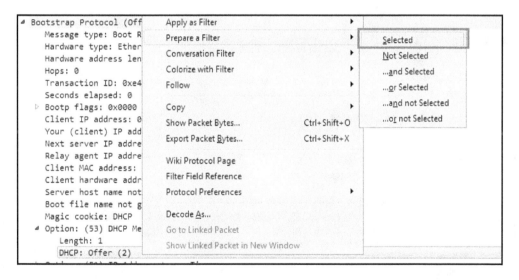

This way, you can filter all of your packet captures based off of just the `Offer`. If we had a whole bunch of `Offers` from a bunch of different servers, that could be a problem.

If we go back to our standard `bootp` filter, you'll see after the `Offer` where the server is now responding to our broadcast it says. I then send out a request for my client. You can see down in the packet details in the options that I'm actually requesting a specific address:

```
⊿ Option: (53) DHCP Message Type (Request)
     Length: 1
     DHCP: Request (3)
⊿ Option: (61) Client identifier
     Length: 7
     Hardware type: Ethernet (0x01)
     Client MAC address: Elitegro_4a:08:12 (f4:4d:30:4a:08:12)
⊿ Option: (50) Requested IP Address
     Length: 4
     Requested IP Address: 192.168.6.76
⊿ Option: (54) DHCP Server Identifier
     Length: 4
     DHCP Server Identifier: 192.168.4.1
⊿ Option: (12) Host Name
     Length: 12
     Host Name: PPMUMCPU0110
⊿ Option: (81) Client Fully Qualified Domain Name
     Length: 28
   ▷ Flags: 0x00
     A-RR result: 0
     PTR-RR result: 0
     Client name: PPMUMCPU0110.packtpub.net
⊿ Option: (60) Vendor class identifier
     Length: 8
     Vendor class identifier: MSFT 5.0
⊿ Option: (55) Parameter Request List
     Length: 12
     Parameter Request List Item: (1) Subnet Mask
     Parameter Request List Item: (15) Domain Name
     Parameter Request List Item: (3) Router
     Parameter Request List Item: (6) Domain Name Server
     Parameter Request List Item: (44) NetBIOS over TCP/IP Name Server
     Parameter Request List Item: (46) NetBIOS over TCP/IP Node Type
     Parameter Request List Item: (47) NetBIOS over TCP/IP Scope
     Parameter Request List Item: (31) Perform Router Discover
     Parameter Request List Item: (33) Static Route
     Parameter Request List Item: (121) Classless Static Route
     Parameter Request List Item: (249) Private/Classless Static Route (Microsoft)
     Parameter Request List Item: (43) Vendor-Specific Information
⊿ Option: (255) End
     Option End: 255
```

This you would not normally see on a new system that just booted up, but because my system, the client, already knew that it had a previous address, it kept that information saved even though it was not configured in the network card. It requested that specific address to renew it and put it back into its configuration. You can see that now it also knows the server. So it sends it out to the server, but doesn't come from a layer 3 IPv4 address because it doesn't have one yet; it's requesting for it. Hence, it still sends it out to the broadcast.

You might see additional options in the preceding screenshot. These option numbers reflect a whole bunch of different things that you can configure in DHCP; it's not just for IP addresses. You'll see this commonly used with voiceover IP phones because you can pass different options, such as option 43, and actually tell it what VLAN it needs to belong to and force it to the other VLAN. You can also tell it where the TFTP server is to retrieve firmware information and all sorts of different things that you can send to a device to automatically configure it. That's why they call it the dynamic host configuration protocol: it's not just for IP. If you want to learn more about it, you can of course take a look at the RFC.

The RFC for DHCP (remember, this is the upgraded version of bootp) is 2131, and you will see that it's quite a lengthy document (http://ietf.org/rfc/rfc2131.txt):

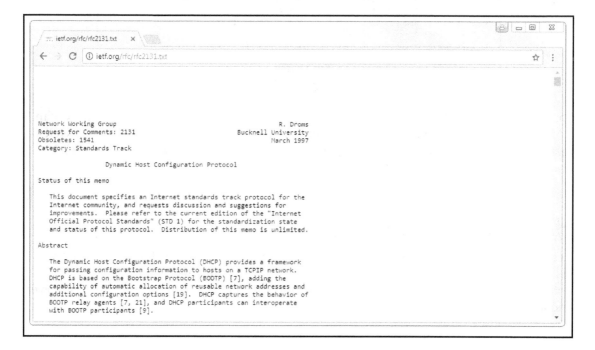

There's quite some information as shown in the preceding screenshot, and it goes through a lot of the functionality of DHCP. Now, it will not include every single option and every single thing that you can configure because some of them are vendor specific; and of course, DHCP has been extended since 1997, with additional add-on features.

We also see packets that say DHCP Inform, and these are requests from my client:

62 1.661	0.073281000	192.168.6.63	255.255.255.255	DHCP	342	DHCP Inform	- Transaction ID 0xb0b66839
125 4.021	0.003302000	192.168.5.242	255.255.255.255	DHCP	342	DHCP Inform	- Transaction ID 0x66dcaae
164 4.793	0.011726000	192.168.7.73	255.255.255.255	DHCP	342	DHCP Inform	- Transaction ID 0x61e19fca
172 5.072	0.056530000	192.168.6.249	255.255.255.255	DHCP	342	DHCP Inform	- Transaction ID 0x53f276fd
194 5.788	0.083492000	192.168.6.46	255.255.255.255	DHCP	342	DHCP Inform	- Transaction ID 0xef2bea3

Now that it has its Layer 3 IPv4 address, it's now requesting additional parameters from the server, and you can see them listed down in the packet details. The following screenshot shows the additional parameters:

```
⊿ Option: (55) Parameter Request List
    Length: 13
    Parameter Request List Item: (1) Subnet Mask
    Parameter Request List Item: (15) Domain Name
    Parameter Request List Item: (3) Router
    Parameter Request List Item: (6) Domain Name Server
    Parameter Request List Item: (44) NetBIOS over TCP/IP Name Server
    Parameter Request List Item: (46) NetBIOS over TCP/IP Node Type
    Parameter Request List Item: (47) NetBIOS over TCP/IP Scope
    Parameter Request List Item: (31) Perform Router Discover
    Parameter Request List Item: (33) Static Route
    Parameter Request List Item: (121) Classless Static Route
    Parameter Request List Item: (249) Private/Classless Static Route (Microsoft)
    Parameter Request List Item: (43) Vendor-Specific Information
    Parameter Request List Item: (252) Private/Proxy autodiscovery
```

So, that's the basics of DHCP. It's a very simple protocol; it's a little bit more complex than DNS but still relatively simple and very useful. There is a IPv6 version of DHCP, as well as many other ways of addressing IPv6 hosts. Up next, we'll take a look at HTTP in *HTTP analysis I*.

HTTP analysis I

In this section, we'll take a look at how HTTP works (what are some of the codes within HTTP and what's inside a packet), source and destination information and some of the options there, and how servers and clients interact and show a connection between a server and a client.

What we'll do is start another packet capture and open up a website. In this example, I opened up a web page to `https://www.npr.org/`, which happens to be an unencrypted website. It uses plain HTTP by default so, that way, the communication is not hidden behind TLS encryption. This way, we can take a look at what actually happens within the HTTP headers.

If we scroll down, we can see we have the `www.npr.org` DNS resolution, our answer, and the beginning of the `SYN, ACK` three-way handshake for the TCP connection:

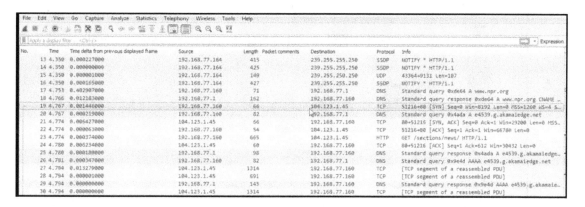

 We will also see some `akamai` DNS resolutions, as well, and that's because if we take a look at `www.npr.org` it is actually hosted off of some `akamai` servers, which is a content distribution network that's distributed around the world, so it's very quick to respond. Hence, it has to resolve some of these additional servers as we go along.

We can see that we have an initial TCP request to the actual server, and then my system asks for `sections /news` because I was opening up the news section on `https://www.npr.org/`:

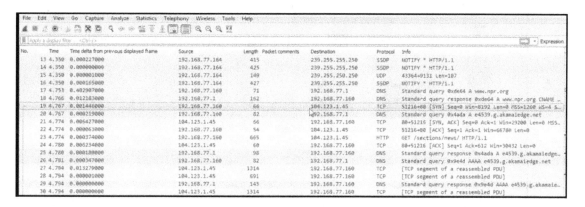

If we scroll down, we'll see that there's an HTTP protocol and some TCP reassembled segment stuff. This is a lot of TCP and I can't see anything for HTTP. Why is that? It's because we have the reassembly enabled in the options. This is something you'll probably want to turn off if you're doing HTTP analysis.

Go to **Edit** | **Preferences...** | **Protocols** | **TCP** and turn off **Allow subdissector to reassemble TCP streams**:

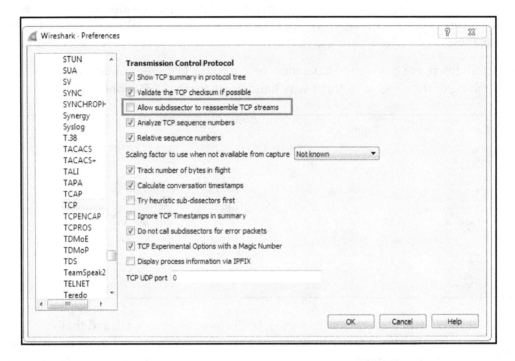

If you turn that off, you can see that we now get some insight into HTTP, and it actually shows up in the **Info** column what are the commands back and forth for the HTTP traffic. You can see that they will now show up properly as HTTP in the **Protocol** column, and it will say that it's a continuation of HTTP as it transmits all of the website information from the server to my client. You can see that the Window size is actually used, as well. We have a nice, big window size and then we have a list of packets that we then acknowledge:

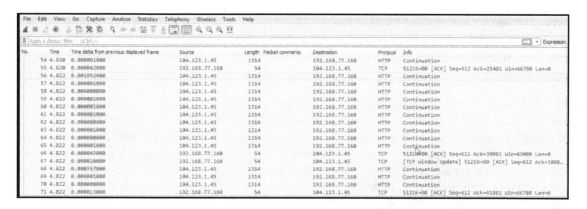

No.	Time	Time delta from previous displayed frame	Source	Length	Packet comments	Destination	Protocol	Info
54	4.820	0.000001000	104.123.1.45	1314		192.168.77.160	HTTP	Continuation
55	4.820	0.000042000	192.168.77.160	54		104.123.1.45	TCP	51216→80 [ACK] Seq=612 Ack=25481 Win=66780 Len=0
56	4.822	0.001952000	104.123.1.45	1314		192.168.77.160	HTTP	Continuation
57	4.822	0.000001000	104.123.1.45	1314		192.168.77.160	HTTP	Continuation
58	4.822	0.000000000	104.123.1.45	1314		192.168.77.160	HTTP	Continuation
59	4.822	0.000001000	104.123.1.45	1314		192.168.77.160	HTTP	Continuation
60	4.822	0.000001000	104.123.1.45	1314		192.168.77.160	HTTP	Continuation
61	4.822	0.000001000	104.123.1.45	1314		192.168.77.160	HTTP	Continuation
62	4.822	0.000000000	104.123.1.45	1314		192.168.77.160	HTTP	Continuation
63	4.822	0.000001000	104.123.1.45	1314		192.168.77.160	HTTP	Continuation
64	4.822	0.000000000	104.123.1.45	1314		192.168.77.160	HTTP	Continuation
65	4.822	0.000001000	104.123.1.45	1314		192.168.77.160	HTTP	Continuation
66	4.822	0.000042000	192.168.77.160	54		104.123.1.45	TCP	51216→80 [ACK] Seq=612 Ack=38081 Win=63000 Len=0
67	4.822	0.000010000	192.168.77.160	54		104.123.1.45	TCP	[TCP Window Update] 51216→80 [ACK] Seq=612 Ack=3808…
68	4.822	0.000357000	104.123.1.45	1314		192.168.77.160	HTTP	Continuation
69	4.822	0.000001000	104.123.1.45	1314		192.168.77.160	HTTP	Continuation
70	4.822	0.000000000	104.123.1.45	1314		192.168.77.160	HTTP	Continuation
71	4.822	0.000013000	192.168.77.160	54		104.123.1.45	TCP	51216→80 [ACK] Seq=612 Ack=41861 Win=66780 Len=0

If we take a look at the HTTP here, my system 77.160 did a GET request. HTTP has two primary commands that we use: GET and POST. A GET request retrieves information while a POST request sends information. So, you know how, in some forms on certain websites or if you make changes to web settings in a profile, you're sending data to the server, telling it to change something on the server: you do that with POST. With GET we are asking for information. So, in this example, I am getting /sections/news, and I'm requesting it over version HTTP 1.1. There is a new version of HTTP, which has recently come in use, and it's based off of Google's SPDY protocol which they had previously created.

If you want to learn more about the SPDY protocol and what it was, you can take a look at that on Wikipedia or on https://www.chromium.org/ —they have a page on this as well. It was really an experimental protocol to speed up traffic, and it has since been deprecated in favor of HTTP 2.0, which has now become the standard. So, the ideas within SPDY have been merged into the HTTP 2.0 standard.

What it did was it basically optimized the HTTP header information and the communications so that it could achieve up to a 50% speed increase on loading a website: that's very powerful and impressive.

So, when you see these GET requests, you'll most likely be seeing 1.1 for some really old clients if you're using, like, a really old program on a very old system, maybe even asking for 1.0. But you may now see 2.0 requests. Probably about a third of the major websites out there now use HTTP 2.0, and these will of course only increase over time. So, we are asking for /sections/news, and then it's insinuated there that we are asking for the index.html page from inside that. Thus, we're asking for a folder structure. By default, HTTP will look for index.html or a couple of other different files indexed at HTML or some other file format. It's the responsibility of the server to serve up that core page that'll first show up.

From the server, we can see that we have a TCP acknowledgment to that GET request, and then the server responds back with: "ok, that sounds good. I will send that to you because I've been able to find that page." Thus, if you request a page that's incorrect, you'll get an error message.

In HTTP, we have different types of commands with different numbers. If you'd like to learn about HTTP in more depth, take a look at the RFC on the IETF website. You're looking for number 2616, and that's for 1.1. Remember I said there's a new one, 2.0, coming out, so of course that's going to have a higher number. When you look through the standard, you'll see a bunch of different codes. You will see a bunch of the code blocks that are available, and the details of each code within it but, if you look to the left, you will see a status code number. Anything that is 200 or 300 are OK. So a 200 OK means "I found the file, no problem." 201 says "ok, I created it." 202 is an accept. These are all good things. If you get a 300, this might be a redirect or to move a file somewhere else. A 400 or a 500 is an error. So, a 400 is a server error. All 400 numbers are server errors saying: "I can't find the file. You're not allowed to get there"; or "It's forbidden"; or "Your method is not allowed"; or "The server is rejecting your request". A 500 error is a client error, so there's a problem on your client side. It is very common to see a 404 error when you try to request a web page that does not exist. You'll see it all over the internet and now everyone's used to it, but a 404 error is: "I cannot find the file." The server says: "I don't know what you're requesting It's, not where you say it is" and it sends back a 404 error.

If we look in the packet details into the HTTP header information you see that we have a server, Apache; this may also say something such as nginx or some other server that's running—Apache is still the most common one. It'll tell you what it's running. If it's running PHP of different versions or Python or something like that, it'll tell you what the content type is. Is it an HTML page? Is it some other kind of content type? Is it an XML page? Sometimes you can have encoding as well. Some pages and some servers allow for compression. So, they'll compress with gzip, which is like creating a ZIP file of the server page that it's sending back so that it is smaller, and so uses less packets and is quicker to send to the client. It takes a little bit of processing power on the server or the clients' sides to do that, but it's usually beneficial. It'll also tell us how long the content length is.

Now that we've gone through all these different protocols, they almost always tell you how long the content is so that you can validate whether you've received everything. We also have an expiry and a cache control. This tells the system how long to save a page. When your client receives the page, it will cache it for a period of time based on this so that it can refer to its local cache if it goes back there again.

So, if you're constantly going back to the same web page all the time, it will load it off of your local cache rather than constantly pull the server for it and use up unnecessary bandwidth and resources on the server. If you don't wish to use the cache, that's when you refresh a page. You can usually press *Ctrl + F5* and it will force the cache in your browser to delete that page and request a new one. If we expand the packet details, we will see **Line-based text data**; it will actually show us the web page itself, as it's sent to us:

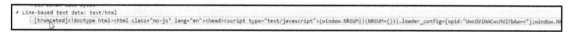

In the next section, we'll dive into HTTP a little bit more, talk about some more problems, and take a look at how you can decrypt TLS-encrypted HTTP data—HTTPS in Wireshark.

HTTP analysis II

We'll investigate some issues with HTTP by looking at the error messages again and how to decrypt HTTPS (which is TLS now) traffic. This also works for SSL.

We'll download an example capture from the **SampleCaptures** section on the Wireshark wiki (`https://wiki.wireshark.org/SampleCaptures`). Once you go to the **SampleCaptures** page, go down to **Specific Protocols and Protocol Families | HyperText Transport Protocol (HTTP)**:

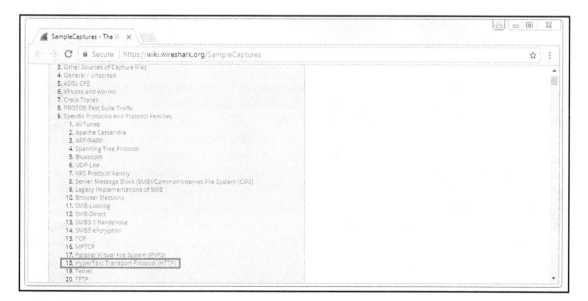

If you click on that, we'll have a list of some basic HTTP captures that we can look at. At the bottom it has a link to **SSL with decryption keys**, and we'll download the top link snake-oil2 070531.tgz file. All you'll need to do is extract that so that you can get to the files within.

By default, Windows can only extract .zip files, so you'll need to download something like 7-Zip or WinRAR in order to open it.

When you open the cap file, you'll see that it's an older file. This is actually from somewhere around 2007, probably, so it's not a pcapng file. But it still opens up just fine in Wireshark. If we look at this, we see that we have a SYN; SYN, ACK; ACK with TCP, so there's our three-way handshake; and then we have an SSL Client Hello, an acknowledgment from the server, then the Server Hello acknowledgment. We start exchanging some key information for creating the SSL encryption; we share the cipher information; then we begin by actually encrypting the data after that. Here, we can see that we have an encrypted handshake and encrypted data. So we have this data that's encrypted, but we can't get to it. We have all this Application Data that you can see, and it is unreadable to us. If we expand any of this in SSL, it's just gibberish. So how do we see the HTTP within it? We need to decrypt it, and with Wireshark you can decrypt SSL or TLS traffic. However, you will need the private key from the server, so if you do not have access to the web server, you cannot do this. This is great if you're on a corporate network where someone's accessing a corporate resource that happens to be encrypted and you want to decrypt it. Well then, you have access to the server so you can retrieve the private key. If you want to do this over the internet, you have to get the private key from whoever's hosting it.

In order to set this up, we'll go to **Edit | Preferences... | Protocols | SSL**. You'll notice that TLS is not listed in the **Protocols** option. TLS falls under the **SSL** because TLS came about after SSL and was built off of SSL; it all falls under the **SSL** protocol section. So, we'll go to configure our RSA keys, the private keys, within the SSL protocol list—even if it is TLS that you're using. So what we'll do is click on **Edit...** and then the plus sign to add a new decryption key. What we'll do is add the IP address of our server. In this example, it was sanitized to 127.0.0.1, which is the localhost; it's just a loopback address. We'll define it as port 443 because we know this is HTTPS traffic that's using standard port 443, and we know that it's http traffic that's behind it. Then, we'll double-click on the **Key File** section and select our key. You can see in that extracted file we have the cap file, a README and the key file. We'll select the key file and then click on **OK**:

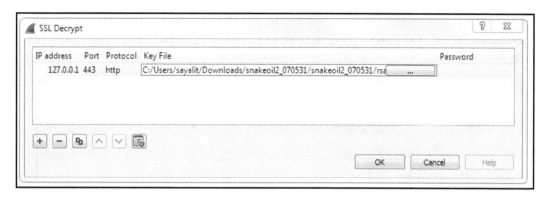

Now you'll see that the data has decrypted itself, so we no longer have all of these TLS protocol listings. It doesn't say `Application Data`; it actually shows us HTTP traffic now. We have now decrypted this. We still have the SSL protocol stuff up because that's actually SSL traffic. It's doing the handshake and exchanging the cipher information so it can encrypt; that's ok. The stuff at the bottom is what we cared about. So now we can actually do our display filter for `http`, and we can filter out just the HTTP communication. We can see that we have few problems in this packet capture. So, we have a `GET` request; it received a response from the server, saying: "I found that file." They then tried to get two more images over HTTP; they were downloading some images: it looks like the Debian logo and something else. Then, there was a response from the server, stating that the resource was not found. Here's that `404` message:

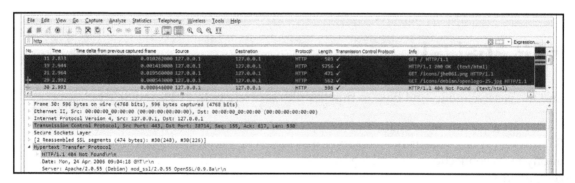

If we go in the packet details, we can see `404` in the HTTP option. You'll also see that it says "expert information". Go down to the bottom left, click on the circle icon, and it will provide you with all of the errors that are in this packet capture:

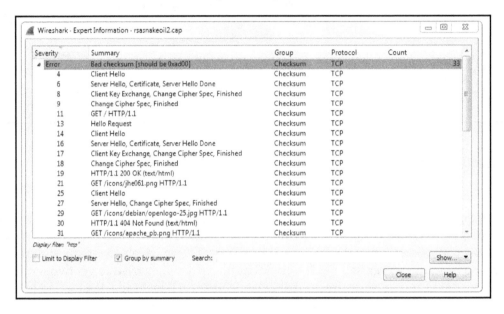

It pulls out HTTP errors along with anything that's of use.

FTP analysis

In this section, we'll take a look at how FTP works, the differences between the active mode and the passive mode, and how to transfer files securely with FPTS and SFTP.

We'll start a capture once again and connect to an FTP server that does not use encryption. So, this server is a Belarus-hosted server that has some Linux ISOs on it, and it allows anonymous connections. I'll just log in with `anonymous`, and we'll use port `21` and click on **Quickconnect**:

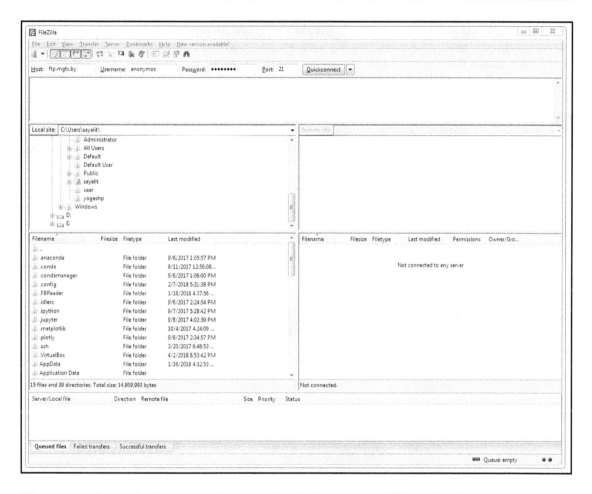

We can see that we've logged in. It states that it does not use TLS, and it lists the root directory:

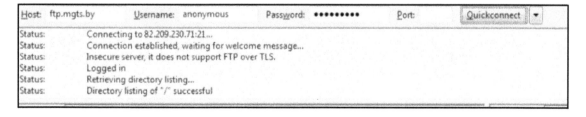

What we'll do is stop that capture, and if we scroll down through the packets we'll see that we have a bunch of other types of traffic here, but then we see some FTP listed. What we can do is create a filter with simply `ftp` in it, and that'll show us all the FTP traffic:

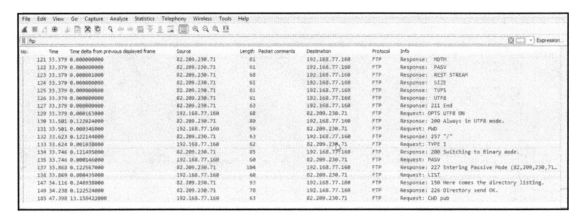

We can see the connection over unencrypted FTP. You can see all the commands, just like with HTTP. If we dig into our **FTP** section in the packet details, we'll see that we have very few commands that are transferred:

```
▲ File Transfer Protocol (FTP)
    ▲ 220 (vsFTPd 2.3.2)\r\n
          Response code: Service ready for new user (220)
          Response arg: (vsFTPd 2.3.2)
```

We connected with our three-way handshake in the packets prior to packet 101. So packets 98, 99, and 100 were the TCP handshake. Then, it states code 220 as being service ready for the new user. So the server says: "I see you've created a connection to me. Send me some authentication". My software asked whether it is capable of using TLS, which is the encryption that we use for HTTPS to allow for encryption. The server responded, saying: "No, I can't do that. Please just log in with the username and password". Then my software, the FileZilla client, said: "Well, what about SSL?" And then the server replied, and said: "No, sorry. I don't do that". "Then my client finally sent it a user, saying: "Here's the user command. I'm going to log in with anonymous". The server accepted that, and said: "Please specify the password".

I provided the password with the PASS command. The server evaluated that and it determined that the credentials were good, and it said: Login successful, code 230. My system then asked what kind of system type it is and what the server's running, and the server responded that it's using a Unix-type server. It then asked for what feature set it has available, and the server responded back with a list of features that it's capable of supporting. Then it says: "That's the end of my list". My client turned on the UTF8 option. The server said: "No problem". Then my client requested **print working directory (PWD)** and the server said: "You are in the root, the / directory. That's the first directory that you're in". My client changed to Binary mode, which is type I. The server responded, saying: "No problem". I then requested passive mode that was another feature that it was capable of providing. Passive mode allows FTP to communicate and transfer its data over a random, dynamic port. So if you were looking at the port numbers here, I'm talking to the server and my client, which is running locally; the client software's running on 52284. It's talking to the server on port 21. The data is just commands, which has to go over another port. With FTP active, the data will transfer over port 20 and all the commands over 21. This is not a common thing to see nowadays because the active mode can easily be hacked. It's not a very secure method because you know exactly what port the data's going to be transferred over. So, if you're in the middle, you know exactly what port to look at in order to capture all of the unencrypted data. Passive mode, on the other hand, allows for a dynamic port to be used, so the data itself will be transferred over a different port. So, it's an enhancement to the FTP protocol. Nowadays, most commonly, you'll see port 21 used for the server commands and then a dynamic port ranging in the thousands for the actual data itself. FTP separates the commands from the data on different port numbers. Not every protocol does that, as we've seen with HTTP, for example, which runs commonly on port 80, and it sends commands and data on the same port, unlike FTP. FTP is a very old protocol.

So we see a response from the server, saying: "We're entering passive mode". I then ask for a list of the directory that I'm in. Remember I asked for what directory I'm under: print working directory. I switched to Binary mode. I said: "Let's use passive. Now give me a list of all the files". The server then says: "ok, I'm sending the directory listing. The directory listing was sent, so this is from the server. And then, I asked to change the directory to a subfolder called pub, and at that point I stopped the capture. What we're missing, though, is the data. You notice I don't have any data. How do I find the data with a dynamic port number using FTP passive mode? You do that by adding an additional filter to your display filter.

You see that we have `ftp`, which is for the commands. If you want the data as well, you'll have to type `ftp || ftp-data`. If you press *Enter*, now we'll get the FTP traffic, as well as the FTP data traffic:

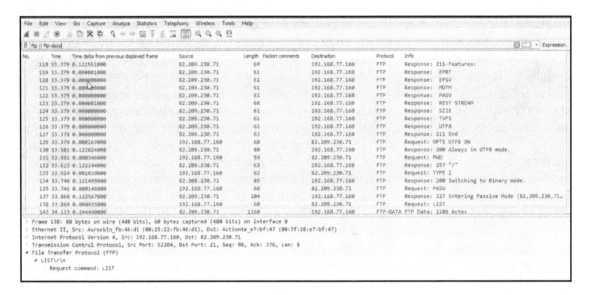

As you can see, my client requested the listing of that directory, the `root` directory; now, the server is sending all of that data over the special port. We can see that the server is running on port `54536`, and it's going to my client running at `52285`. So that's the difference between FTP active and FTP passive.

Now, there's also FTPS and SFTP. If you notice, all of the packets are unencrypted. We see all the commands—the login username and password. This is not good for going over the public internet in a nonanonymous mode. If you're using a username and password, you should be encrypting that—you should be encrypting your data. You can do that using SSL or TLS, just like I mentioned with HTTP as we've seen before, and many other protocols that you can encrypt with TLS. So, FTPS is FTP encrypted with TLS or SSL. That's the more standard method of doing that using the FTP protocol as it's already implemented, but then encrypting on top of it.

There's also **Secure Shell File Transfer Protocol** (**SFTP**). SFTP runs over a different port; it runs over port 22. So, you'll actually want to filter based on `ssh` in that case because the FTP traffic is traveling over SSH, the secure shell port; so, you could use `ssh` as your filter. Now obviously that's not here in this capture, but that works as the filter you will use, and all the traffic will go over that. But it would all be encrypted—just like with FTPS, which is encrypted—you will need the server private key in order to decrypt it, like you saw in the previous section with HTTPS. Note that there are two protocols for transferring files in a secure manner, and they are different. FTPS uses the standard FTP but encrypts it with TLS or SSL. SFTP is its own animal; it has its own protocol and application. SFTP is completely separate, and it runs file transfers over SSH.

In this section, we discussed DHCP: how that protocol functions and how to take a look at some of the options that are in it.

Summary

We looked at HTTP, both in an unencrypted fashion and an encrypted fashion, and how to decrypt it. We also talked about FTP in all of its many flavors: active mode, passive mode, and the encrypted flavors of FTPS and SFTP.

In `Chapter 9`, *Application Protocol Analysis II*, we'll continue with some additional application protocols that you will run into.

Application Protocol Analysis II

9

In this chapter, we'll cover the following topics:

- Email analysis, including POP and SMTP
- 802.11 (or wireless and Wi-Fi) analysis
- VoIP analysis for voice over IP telephony
- VoIP play back of the captured traffic to be able to hear the issues that may be occurring in it

Email analysis

In this section, we will take a look at what POP and SMTP are used for, as well as investigating the communication in POP and SMTP and learning what some of the codes are; whether they're successful, or error codes, or something like that.

Now, we have three main protocols that we use for email on the internet:

1. POP
2. SMTP
3. **Internet Message Access Protocol (IMAP)**

These three are used by clients such as mobile devices or software running on a computer; some sort of standalone application such as Thunderbird or Outlook. There are other protocols as well if you're dealing with Outlook for exchange, for example, but primarily, if you're using a generic application and you're accessing your email server in a generic fashion, you'll be using one of these three protocols.

We'll be focusing on POP and SMTP.

POP and SMTP

POP is used for retrieving emails from a mail server. SMTP is used for sending emails to a mail server and sending emails between mail servers. If you'd like to learn more about POP and SMTP, take a look at these RFCs:

- RFC 1939 for POP, which you can see in the following screenshot (`https://tools.ietf.org/html/rfc1939`):

- RFC 2821 for SMTP, which you can see in the following screenshot (`https://tools.ietf.org/html/rfc2821`):

In the preceding screenshot, you'll see there are a number of commands that are transmitted between the client and the server in order to convey what they want to accomplish in the connection. Just like we've seen with HTTP and FTP, and some other protocols, there's an agreed-upon language that is used in order to execute certain things. We'll take a look at that in a packet capture.

Now, it's very common in modern use to encrypt your data, especially email nowadays, so POP is often encrypted with SSL or TLS. You also do the same thing with SMTP or IMAP: you can now encrypt all these protocols with SSL or TLS. This is an unencrypted communication so that we can take a look at all the commands that are passed back and forth.

What we see in the packet capture is we have in the beginning a three-way handshake for TCP; POP is transmitted over TCP. And then we have our POP communication as well as some acknowledgments, and then the `FIN` closure.

There is a filter for POP. You can right-click on your **Protocol** and apply it as a filter; or you can simply enter `pop`, and that will filter your traffic:

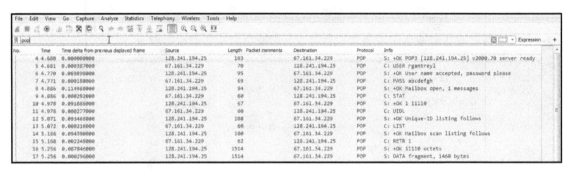

However, if you notice, we have lost the handshake information at the beginning and the end, and some of the acknowledgment packets that were there earlier in between the data have now been removed because they weren't part of POP. They were TCP acknowledgments and handshake. So it might be better if you determine the addresses that are in use or the port that is in use, and end up filtering on that instead of `pop` as the display filter. So what we have is our three-way handshake. We have our client and server. The client is requesting a connection. The server says: ok, no problem. The client then says: ok, I acknowledge a connection; and we have a three-way handshake. After that the server then responds and says: ok, we are connected with POP3. The server is ready; you can see that in the packet details.

If we expand our **Post Office Protocol**, we'll have some more information. It says that the server's ready:

```
▷ Transmission Control Protocol, Src Port: 110, Dst Port: 1643, Seq: 1, Ack: 1, Len: 49
◢ Post Office Protocol
    +OK POP3 [128.241.194.25] v2000.70 server ready\r\n
```

So that's a good message. The client then passes its user, and you see that since this is unencrypted, the user is in plain text:

4 4.680	4.593677000	128.241.194.25	103	67.161.34.229	POP	S: +OK POP3 [128.241.194.25] v2000.70 server ready
5 4.681	0.000387000	67.161.34.229	70	128.241.194.25	POP	C: USER rgantreyl
6 4.770	0.089890000	128.241.194.25	95	67.161.34.229	POP	S: +OK User name accepted, password please
7 4.771	0.000180000	67.161.34.229	69	128.241.194.25	POP	C: PASS abcdefgh
8 4.886	0.114988000	128.241.194.25	84	67.161.34.229	POP	S: +OK Mailbox open, 1 messages
9 4.886	0.000292000	67.161.34.229	60	128.241.194.25	POP	C: STAT
10 4.978	0.091886000	128.241.194.25	67	67.161.34.229	POP	S: +OK 1 11110
11 4.978	0.000277000	67.161.34.229	60	128.241.194.25	POP	C: UIDL
12 5.071	0.093466000	128.241.194.25	108	67.161.34.229	POP	S: +OK Unique-ID listing follows
13 5.072	0.000216000	67.161.34.229	60	128.241.194.25	POP	C: LIST
14 5.166	0.094390000	128.241.194.25	100	67.161.34.229	POP	S: +OK Mailbox scan listing follows
15 5.168	0.002249000	67.161.34.229	62	128.241.194.25	POP	C: RETR 1
16 5.256	0.087846000	128.241.194.25	1514	67.161.34.229	POP	S: +OK 11110 octets

We passed the USER command to the server. The server responds and says: ok, the username is good. Please send me a password. We then send a password. Again, this is unencrypted, so it's in clear text. And you can see it's very easy to determine what's going on. We have the commands, which are named USER, PASS, and OK. The server responds and says: ok, mailbox has been opened up. Thank you for your credentials. You have one message that is unread. The client then asks the server for some status; the server responds with the status message. We then get a unique ID listing, and then the client asks for a list. Highlighted in the preceding screenshot is a list of whatever's in the mailbox. The mailbox scan is completed, and it sends us the data, which is the number of messages and the number of bytes. The client then says: all right, let's retrieve message number 1. The server then says: ok, these are the octets that you requested.

And you can see that down in the packet details; we have not only the number of octets but we have the actual email itself, which includes the header information; the sent and received information; the From; the To; the Subject; the Date; and then the actual data inside the email itself:

```
Post Office Protocol
 ⊿ +OK 11110 octets\r\n
     Response indicator: +OK
     Response description: 11110 octets
 Return-Path: bbelch@packet-level.com\r\n
 Received: from mx20.stngva01.us.mxservers.net (204.202.242.7)\r\n
 \tby mail11d.verio-web.com (RS ver 1.0.95vs) with SMTP id 3-0575327743\r\n
 \tfor <rgantreyl@packet-level.com>; Mon, 15 Jan 2007 16:49:06 -0500 (EST)\r\n
 Received: from mxw1100.verio-web.com [161.88.148.09] (EHLO GIGA)\r\n
 \tby mx20.stngva01.us.mxservers.net (mxl_mta-1.3.8-10p4) with ESMTP id d05fba54.2509.132.mx20.stngva01.us.mxservers.net:\r\n
```

As you can see, the data then continues, as shown in the following screenshot:

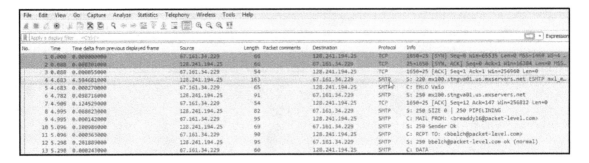

16 5.256	0.087846000	128.241.194.25	1514	67.161.34.229	POP	S: +OK 11110 octets
17 5.256	0.000296000	128.241.194.25	1514	67.161.34.229	POP	S: DATA fragment, 1460 bytes
18 5.257	0.000025000	67.161.34.229	54	128.241.194.25	TCP	1643→110 [ACK] Seq=58 Ack=3154 Win=256960 Len=0
19 5.257	0.000240000	128.241.194.25	1230	67.161.34.229	POP	S: DATA fragment, 1176 bytes
20 5.258	0.001077000	128.241.194.25	1514	67.161.34.229	POP	S: DATA fragment, 1460 bytes
21 5.258	0.000017000	67.161.34.229	54	128.241.194.25	TCP	1643→110 [ACK] Seq=58 Ack=5790 Win=256960 Len=0
22 5.258	0.000302000	128.241.194.25	1514	67.161.34.229	POP	S: DATA fragment, 1460 bytes
23 5.259	0.000931000	128.241.194.25	1514	67.161.34.229	POP	S: DATA fragment, 1460 bytes
24 5.259	0.000023000	67.161.34.229	54	128.241.194.25	TCP	1643→110 [ACK] Seq=58 Ack=8710 Win=256960 Len=0
25 5.343	0.083831000	128.241.194.25	338	67.161.34.229	POP	S: DATA fragment, 284 bytes
26 5.351	0.007775000	128.241.194.25	1514	67.161.34.229	POP	S: DATA fragment, 1460 bytes
27 5.351	0.000039000	67.161.34.229	54	128.241.194.25	TCP	1643→110 [ACK] Seq=58 Ack=10454 Win=256960 Len=0
28 5.437	0.086159000	128.241.194.25	966	67.161.34.229	POP	S: DATA fragment, 912 bytes
29 5.562	0.125231000	67.161.34.229	54	128.241.194.25	TCP	1643→110 [ACK] Seq=58 Ack=11366 Win=256048 Len=0
30 8.371	2.809058000	67.161.34.229	62	128.241.194.25	POP	C: DELE 1
31 8.460	0.088840000	128.241.194.25	75	67.161.34.229	POP	S: +OK Message deleted

So, the first packet was basically the header information with the beginning of the data, and then the data continues. We then acknowledge the packets. We keep retrieving data, then the client says: delete the message. We have downloaded the email; now delete the message. This is an option that is changeable.

You can tell your client to leave a message on the server, but traditionally it is done so that you download it locally and you would delete it off the server.

Nowadays, with our web-based email, or if you're using Gmail or Yahoo! or something like that, we now most often leave everything on the server and have it archived there. But this is a very old protocol, and so it was based on local storage.

The server responds saying that it deleted the message, and then we quit the connection and close out.

It says: ok, sayonara. Then finally, we close out the TCP connection with the FIN and ACK series.

Now, let's take a look at SMTP. SMTP is used to transfer email between a client and a server in order to send it to a routed recipient:

No.	Time	Time delta from previous displayed frame	Source	Length	Packet comments	Destination	Protocol	Info
1 0.000	0.000000000		67.161.34.229	66		128.241.194.25	TCP	1650→25 [SYN] Seq=0 Win=65535 Len=0 MSS=1460 WS=4
2 0.088	0.088301000		128.241.194.25	66		67.161.34.229	TCP	25→1650 [SYN, ACK] Seq=0 Ack=1 Win=16384 Len=0 MSS
3 0.088	0.000055000		67.161.34.229	54		128.241.194.25	TCP	1650→25 [ACK] Seq=1 Ack=1 Win=256960 Len=0
4 4.683	4.594681000		128.241.194.25	163		67.161.34.229	SMTP	S: 220 mx100.stngva01.us.mxservers.net ESMTP mx1_m...
5 4.683	0.000270000		67.161.34.229	65		128.241.194.25	SMTP	C: EHLO Vaio
6 4.782	0.098716000		128.241.194.25	91		67.161.34.229	SMTP	S: 250 mx100.stngva01.us.mxservers.net
7 4.906	0.124529000		67.161.34.229	54		128.241.194.25	TCP	1650→25 [ACK] Seq=12 Ack=147 Win=256812 Len=0
8 4.995	0.088823000		128.241.194.25	82		67.161.34.229	SMTP	S: 250 SIZE 0	250 PIPELINING
9 4.995	0.000142000		67.161.34.229	95		128.241.194.25	SMTP	C: MAIL FROM: <breaddy16@packet-level.com>
10 5.096	0.100986000		128.241.194.25	69		67.161.34.229	SMTP	S: 250 Sender Ok
11 5.096	0.000363000		67.161.34.229	90		128.241.194.25	SMTP	C: RCPT TO: <bbelch@packet-level.com>
12 5.298	0.201889000		128.241.194.25	95		67.161.34.229	SMTP	S: 250 bbelch@packet-level.com ok (normal)
13 5.298	0.000243000		67.161.34.229	60		128.241.194.25	SMTP	C: DATA

What we see here is a series of SYN and ACK of a user creating a TCP connection for SMTP, and then we see the SMTP response from the server. And of course, we have SMTP listed in the **Protocol** column. We could filter on that by simply typing `smtp`:

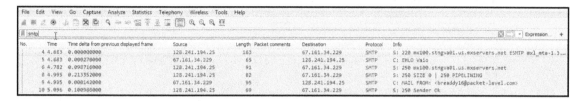

Just like with `pop`, we'll lose some of the information there from TCP, so it might be better to do your filter based on `ip` and `port`.

Well, what we see in the previous screenshot, after the three-way handshake from the server, the 128 address, is that it responds with a 220:

```
▲ Simple Mail Transfer Protocol
    ▲ Response: 220 mx100.stngva01.us.mxservers.net ESMTP mxl_mta-1.3.8-10p4; Mon, 15 Jan 2007 16:49:50 -0500 (EST); NO UCE\r\n
        Response code: <domain> Service ready (220)
        Response parameter: mx100.stngva01.us.mxservers.net ESMTP mxl_mta-1.3.8-10p4; Mon, 15 Jan 2007 16:49:50 -0500 (EST); NO UCE
```

A 220 response code is a service ready, which means everything is good. This will look familiar to POP and HTTP.

Many of these protocols, especially the ones that are older, use these different numbered response codes. Just like with the HTTP, 200 series response codes are good. So we see a 220, everything's good to go. We also see that we have ESTMP. This is for the enhanced version of SMTP. Just like with POP and FTP, SMTP is a very old protocol, and it's been extended and enhanced over the years. There's a newer version of SMTP, which is transmitted using new commands, with an E in the front:

No.	Time	Time delta	Source	Length	Destination	Protocol	Info
4	4.683	4.594681000	128.241.194.25	163	67.161.34.229	SMTP	S: 220 mx100.stngva01.us.mxservers.net ESMTP mxl_m...
5	4.683	0.000270000	67.161.34.229	65	128.241.194.25	SMTP	C: EHLO Vaio
6	4.782	0.098716000	128.241.194.25	91	67.161.34.229	SMTP	S: 250 mx100.stngva01.us.mxservers.net
7	4.906	0.124529000	67.161.34.229	54	128.241.194.25	TCP	1650→25 [ACK] Seq=12 Ack=147 Win=256812 Len=0
8	4.995	0.088823000	128.241.194.25	82	67.161.34.229	SMTP	S: 250 SIZE 0	250 PIPELINING

The client then sends an EHLO, which is the enhanced version of HELO for the traditional original protocol, for it to create a connection. The server responds to the HELO request and creates a connection. We then acknowledge that, and then the server responds with a listing of what it can do. There are some features that it has, and we see that it has **PIPELINING**:

```
⊿ Simple Mail Transfer Protocol
    ⊿ Response: 250-SIZE 0\r\n
        Response code: Requested mail action okay, completed (250)
        Response parameter: SIZE 0
    ⊿ Response: 250 PIPELINING\r\n
        Response code: Requested mail action okay, completed (250)
        Response parameter: PIPELINING
```

PIPELINING is an option in the server that says that it can accept multiple commands without having to wait for each one, so our client can then send a number of things at once and it doesn't have to wait. Our client then says that it'll create an email message; and if you notice, it says MAIL FROM:

```
 9 4.995  0.000142000        67.161.34.229     95      128.241.194.25    SMTP   C: MAIL FROM: <breaddy16@packet-level.com>
10 5.096  0.100986000        128.241.194.25    69      67.161.34.229     SMTP   S: 250 Sender Ok
11 5.096  0.000363000        67.161.34.229     90      128.241.194.25    SMTP   C: RCPT TO: <bbelch@packet-level.com>
12 5.298  0.201889000        128.241.194.25    95      67.161.34.229     SMTP   S: 250 bbelch@packet-level.com ok (normal)
```

Actually, remember when we looked at the POP message, and there was a From and a To and a Subject, and the actual body field? Exactly what you saw in the POP request, you see here in the SMTP. So, we have the prepended command of MAIL, but you see From and the email address, just like you would when you open it up in your client software after you pulled the message. We are literally writing an email with commands. The email itself is not like a data package that is just bundled up into a little file and sent to the server. This is old enough that we are actually crafting the email command by command and line by line in SMTP. So we are saying: We're creating an email and it's from the following person. The server says: ok, looks good. RCPT To says that so we are sending it to the following person. The server says: ok. We then say: here's some data (the data being the body of our message):

```
13 5.298  0.000243000        67.161.34.229     60      128.241.194.25    SMTP   C: DATA
14 5.385  0.086553000        128.241.194.25    100     67.161.34.229     SMTP   S: 354 Start mail input; end with <CRLF>.<CRLF>
15 5.396  0.011263000        67.161.34.229     1514    128.241.194.25    SMTP   C: DATA fragment, 1460 bytes
16 5.396  0.000025000        67.161.34.229     1514    128.241.194.25    SMTP   C: DATA fragment, 1460 bytes
17 5.396  0.000013000        67.161.34.229     1476    128.241.194.25    SMTP   C: DATA fragment, 1422 bytes
18 5.496  0.099800000        128.241.194.25    60      67.161.34.229     TCP    25→1650 [ACK] Seq=277 Ack=3015 Win=65535 Len=0
19 5.496  0.000022000        67.161.34.229     59      128.241.194.25    IMF    from: "Brian Readdy16" <breaddy16@packet-level.com...
20 5.500  0.004146000        128.241.194.25    60      67.161.34.229     TCP    25→1650 [ACK] Seq=277 Ack=4437 Win=64113 Len=0
21 5.718  0.217512000        128.241.194.25    60      67.161.34.229     TCP    25→1650 [ACK] Seq=277 Ack=4442 Win=64108 Len=0
22 6.360  0.642102000        128.241.194.25    102     67.161.34.229     SMTP   S: 250 0-0484658135 Message accepted for delivery
```

The server responds and says: ok, and let me know when you're done with the email message by putting the following commands at the end of your message. Then, the client provides a number of packets here of the actual email data that it wants to have in the body. See some additional packets related to that? We have some acknowledgments from the server for some of these data packets, then a response from the server saying that the message was accepted and it's going to send it to a recipient.

We acknowledge that and tell it: ok, I'm done; and we quit the SMTP connection:

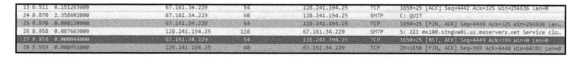

Then we finish out the TCP connection with some FINs and ACKs; we have some explicit resets in the end as well, and that is the termination of our connection.

802.11 analysis

In this section, we'll take a look at wireless connection issues and how they look in Wireshark.

The `802.11` standard has been around for quite some time. You can find more info at `http://ieee802.org/11/Reports/802.11_Timelines.htm`:

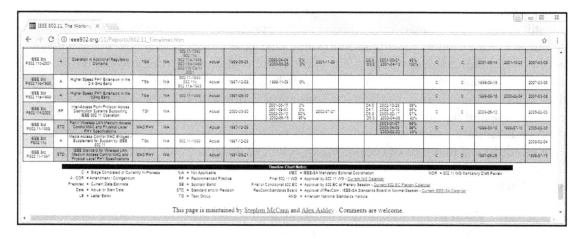

It originally started in 1997. You can see in the preceding screenshot that this was the year it was ratified. It actually began in 1991. Since then it has had many changes to it, including `802.11a`, which some people may remember. We have `802.11b`, `802.11g`, and so on. As we go up in time, you can see how many different flavors of `802.11` there are out there. Now, not all of them have been used for normal home networks or office networks; some of these are specialty versions for long distance or low power such as WiMAX. You can see that all the ones that I've been going through have been superseded by a newer version, and these newer versions are at the top:

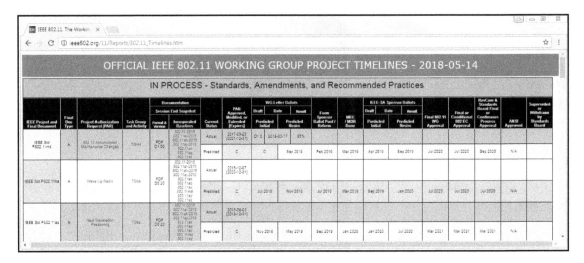

These are the ratified standards. You can see that even though `802.11a` and b and g and n are still supported by hardware out there, still supported by operating systems, the actual standard is, for example, for this version we're talking about, `802.11ac`. So `ac` actually encompasses the `ac` standard, the `n`, the `g`, the `b`, the `a`, and so on. And you can see there are additional versions of `802.11` that are here, such as the one that'll be used for what's called the `TV White Space` area. So these are sections of bandwidth that are available for potential use and reprioritization by the FCC. We have some additional ones, which are proposed standards, such as `ah` and `ai`. Then, there are all these other ones out there that give us additional methods of connecting over longer distances and greater speeds, and things like that. All of this is available on the IEEE website, and you can learn much more about it at `http://ieee802.org/`.

Packt has a number of wireless-related books and video courses that you can take a look at and continue your education in wireless networking.

The following is an example of a packet capture done by a device that could capture the `802.11` frames and provide them to Wireshark:

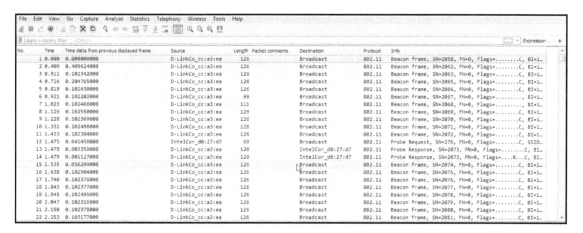

This is commonly done with a wireless card that is capable of not only enabling promiscuous mode but monitor mode in order to view all of the data that we want for all of the different channels that we use. Additionally, there's spectrum analyzers out there such as Wi-Spy and others that are dongles that you can attach to a laptop or something like that; you can go around and investigate the spectrum analysis side of things.

So we're looking at packet captures, but remember there's more to it than just the packet captures. Wireless brings in a whole slew of additional issues such as signal strength and interference with other devices. But looking at the actual data, it shows as essentially Ethernet. I looks like Ethernet to us when we capture it, and that's done on purpose; the `802.11` standard is that it's supposed to look like Ethernet, but over a wireless connection instead of a wired connection. That's to make our life easier and make it more cross-compatible.

And what we see here in the capture is we have the **Protocol** listed as `802.11`, so Wireshark knows that it's an `802.11` capture. This is an `802.11` frame, and it's a `Beacon frame`:

```
IEEE 802.11 Beacon frame, Flags: ........C
    Type/Subtype: Beacon frame (0x0008)
  ▷ Frame Control Field: 0x8000
    .000 0000 0000 0000 = Duration: 0 microseconds
    Receiver address: Broadcast (ff:ff:ff:ff:ff:ff)
    Destination address: Broadcast (ff:ff:ff:ff:ff:ff)
    Transmitter address: D-LinkCo_cc:a3:ea (00:13:46:cc:a3:ea)
    Source address: D-LinkCo_cc:a3:ea (00:13:46:cc:a3:ea)
```

We see there's `Beacon frame` and there are some flags. We have a `Frame Control Field`. We have the `Receiver` and `Destination`; it looks very similar to what you'd expect on layer 2 on a wired network. You have a from and a to MAC address, and you've got some additional information: some additional flags, and so on. Now, of course, there are some other options in here for it to work on wireless, such as our `BSS Id`: what device we are connected to, what access point we are connected to; and what kind of frame this is, because there are different kinds of frames in `802.11`:

```
Source address: D-LinkCo_cc:a3:ea (00:13:46:cc:a3:ea)
BSS Id: D-LinkCo_cc:a3:ea (00:13:46:cc:a3:ea)
.... .... .... 0000 = Fragment number: 0
1000 0000 1010 .... = Sequence number: 2058
Frame check sequence: 0x5fce156d [correct]
[FCS Status: Good]
```

We, of course, have our Frame check sequence, just like we had with the standard Ethernet.

Now, what is a beacon frame? A beacon frame is transmitted normally every 100 milliseconds by an access point in order for it to declare to all of the listening devices that it is providing the following network; that it is beaconing; and that it can support the following network name. You will see a lot of beacon frames in a wireless capture. They occur constantly. If you do not see them constantly, then that is a potential issue. Now, if you want to filter, or if you have a packet capture that includes other erroneous packets in it and other frames, or if you want to filter only your wireless traffic, you can do the `wlan` filter:

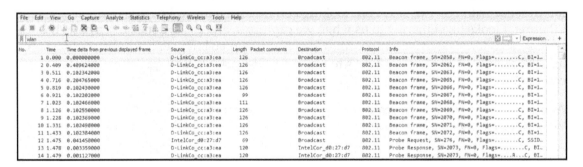

If you use `wlan`, that will include all of the `802.11` protocol frames. Something else that you might also want to do (leave `wlan` filter on) is not show beacon frames. Let's say that you know for a fact after looking through things that beacon frames are consistent and everything is good with them; then we don't need to worry about them.

What we'll do is right-click on the **Beacon frame** in the packet details and go to **Apply as Filter** | **Not Selected**:

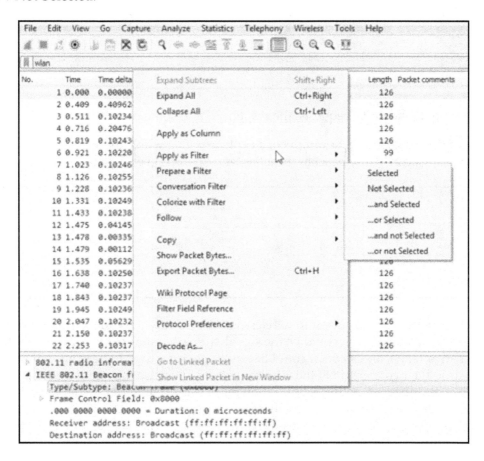

This way, we are selecting all of the wireless LAN frames, but not anything that is a beacon. And that really trims down our capture:

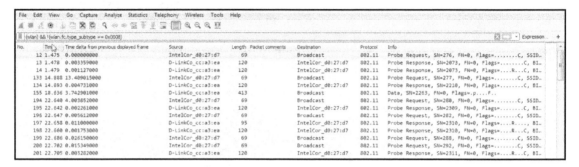

Now, we see only the probe requests, the responses, and some data.

Speaking of beacons, here's an example of a capture with a whole lot of beacons. Everything is just constantly beacons, beacons, and beacons everywhere:

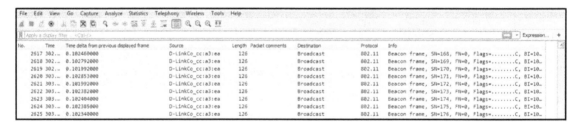

A common problem that can occur in wireless is a drop in beacons due to some sort of problem on an access point. It could be the signal-strength issue or the firmware on the device, a bad antenna, and so on. It could be anything. But a drop in beacons will cause clients to drop their connection to the wireless network because they think the wireless network has disappeared. So the capture shown in the preceding screenshot includes a problem with beacons, but we don't really see anything pop up that's obvious. Nothing's bright red, popping up to us; there are no changes really; it's just all consistent. So how do we see the problem?

Remember back to the section on statistics. Now, go to **Statistics** | **I/O Graph** on our beacons:

We can see that we absolutely have a drop that shows us we've got a problem with our access point:

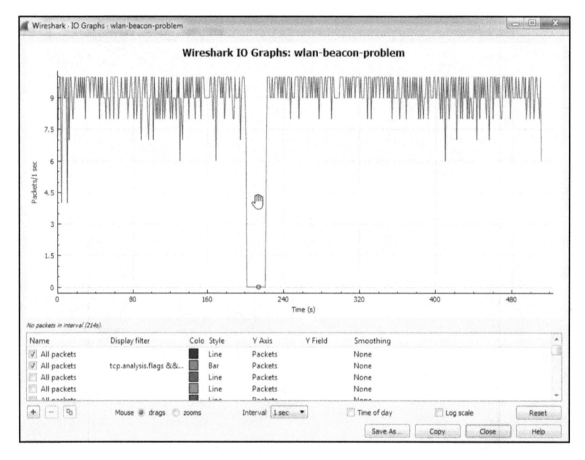

Make sure you certainly use your graphing capabilities when doing wireless troubleshooting because there are a lot of variables that are not controllable by us due to the nature of it transmitting over radio waves. We need to do a lot more visualization in order to see what's going on, especially when you have thousands and thousands of frames like this and there's a problem in there, but you don't know where, and exactly to what extent. Using graphing can be a very big help for you.

Additionally, we have a capture that shows a signal issue:

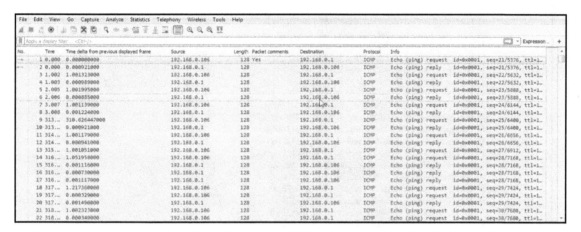

So, we'll do a `ping`. We have `(ping) request`, `(ping) reply`, and so on. Now just scroll down through the capture and see if you find anything that's odd or an obvious problem.

As we scroll through, we can see all `Echo (pings)`: request and reply repeatedly. Do you see any problem? Well there is a very obvious problem.

As we scroll through, look at the fact that we have so many requests and very few replies. That is a very blatant problem. If we go into the expert information, we can see in the warnings that it says **No response seen to ICMP requests**:

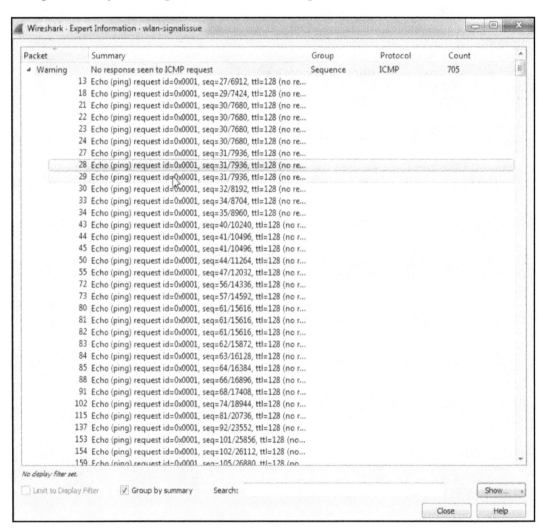

Of course, you know that as we click through each request, it'll show us each packet along the way. We have all these requests that did not receive a reply, but you don't see any other problems in the capture.

Now remember, ICMP does not offer any sort of stateful connection like TCP does, for example, so we don't get any responses that indicate that there are really any problems. We just make an attempt, it fails, and we just keep attempting over and over again. So being able to take a look at what's in the capture, even just scrolling through it and taking a look at what the pattern looks like, can be useful because we don't really have much else to go on.

Now of course, you could also graph this out and have two graphs, one that requests and one that replies, and you can see how they may not match up in the I/O graph. This is indicative of a signal issue, so what we're looking at here is we have some data packets that are making it and some that aren't. However, in the capture, it'll not tell us the signal strength anywhere. We have to determine what the problem is based on the information we have, and from looking at this it looks like a signal problem because we have packet loss. Now, on the testing device, you'll most likely see that in your command window it'll say that so many packets have been lost during the following ping attempts. In the drivers, you may see that the wireless signal is low. There are other ways you can look at this, but if you're looking at just a pure packet capture of ICMP, this is an example of what the problem would look like, such as a low signal issue.

Wireless is a vast topic that is very in-depth, and I highly recommend you spend time going through a full wireless course or a series of books to learn it if you are going to end up supporting that and that's something you're interested in doing.

A short video or book on how to look at it in Wireshark and what certain things might look like in a Wireshark capture is useful, but there's much more going on with wireless because we're using radio waves and signal strengths and such. Thus, it's highly recommended to go out and learn more about wireless in order to understand how things really work behind the scenes.

VoIP analysis

In this section, we'll take a look at how SIP works when a connection is created between two phones, and how RTP works to transmit the live data between the two.

The example capture that we'll use for this chapter and the next one is available on the Wireshark SampleCaptures page (`https://wiki.wireshark.org/SampleCaptures`). If you scroll down and look for the **SIP and RTP** section, we'll be using the **MagicJack+ short test call**:

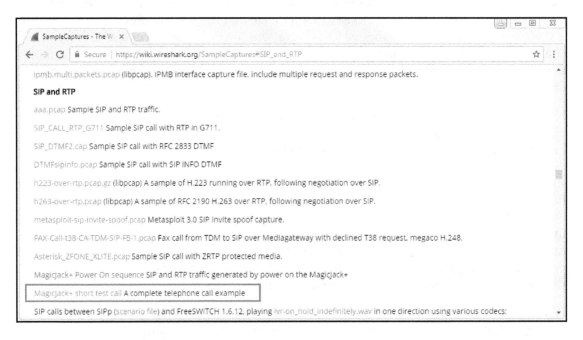

Download **MagicJack+ short test call** and open it in Wireshark.

Once you have that open, we'll take a look at our capture and notice that we have a variety of packets; it has not been yet been cleaned up:

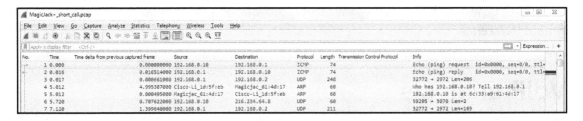

We can see that there's some ARP, some UDP traffic, ICMP, some SIP, and some RTP; we also have some SMB in the end. So there's a mixture of stuff; this is like a real capture. In order to pick out just the SIP traffic, which is one of the many protocols but the most common protocol to use for VoIP, we simply enter `sip` for our display filter and apply it:

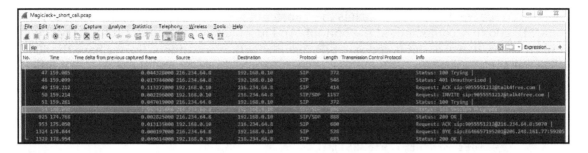

There are a number of other protocols in use as well, such as Skinny, which is a very popular one for Cisco networks, but the standard one is SIP. What SIP does is create the connection between two devices. If you look at the left-hand side of the following screenshot, you can see the **Time** column. We have 159 seconds. So, what we have here is one series of connection attempts. It's all very quick, very much in the same time frame. Then, we have another packet: 166 seconds, and then we have another couple in the 170s. It's important to note because we're missing some data.

As mentioned earlier, the capture also consists of some RTP traffic. What happens is, SIP creates the connection between the two devices. However, the actual data, that is, the video traffic or the voice traffic (whatever it is), is transmitted over the real-time protocol, RTP, directly between the two devices. So if we're filtering on SIP, we'll not be able to see it. What we need to do is add rtp to this and that'll flush out some of this missing time frame we have here.

For that, we'll type sip || rtp, and you can see that we now have our connection creation with SIP:

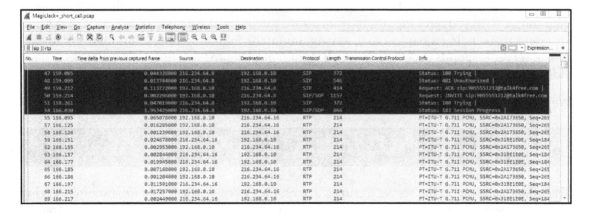

As you can see, we have RTP, which is the transfer of the voice traffic, and then we have some SIP goodbye commands at the bottom. And just like with the other protocols we've spoken about, they use status codes in SIP, just like with many others. So, we have the 400 series, which is a problem; we have 100 series, which is a success; and there are some other commands as well.

 If you would like to learn more about SIP, take a look at RFC 3261, which gives you a great breakdown as to how SIP functions (`https://tools.ietf.org/html/rfc3261`).

What we see in the preceding capture are a number of connections and then some actual data traffic sent. And this is actually a very good capture because it shows us a problem as well as a solution.

So, earlier, we discussed about status codes. Here, we have a `Status: 100 Trying`:

What we have is a local device, which looks like some sort of phone, that is trying to create an SIP connection out to magicJack. It's providing an `INVITE` command, saying: Please invite the following device into the connection. And this is being sent to `216.234.64.8` address; it's being sent to the magicJack SIP server. The SIP server is handling the connection between one phone and the other, and then once that connection is established it is hands off with the communication. It does not act as a proxy and pass through all the data traffic, it only creates the connection. So we can see that here we have `216.234.64.8`. Our device is sending a request to it, saying: Invite the following phone number. Try to call it. The server responds back and says: All right. I'm going to try it out, but you're not authorized.

We then have a second request:

The phone calls out to the server again and says: Please invite the following number. Trying to call it. The server responds back, saying: ok. I'm trying it. I'm going to ring that phone. We then have a success, and it provides us with a `Session Progress`.

If we expand SIP, we can see that we have **Session Progress**:

```
⊿ Session Initiation Protocol (183)
    ⊿ Status-Line: SIP/2.0 183 Session Progress
        Status-Code: 183
        [Resent Packet: False]
        [Request Frame: 50]
        [Response Time (ms): 6816]
    ▷ Message Header
    ▷ Message Body
```

So, we have a success here.

Then, we have additional data about who the call is from and to, as well as what's in the body:

```
⊿ Message Header
  ▷ Via: SIP/2.0/UDP 192.168.0.10:59205;branch=z9hG4bKc0a8000a052182706faf2cbf3d;rport=59205;received=206.248.161.77
  ▷ Contact: <sip:4165551212@216.234.64.8:5070>
  ▷ To: <sip:9055551212@talk4free.com>;tag=30da0aed-co12170-INS015
  ▷ From: "unknown"<sip:E646657195201@talk4free.com>;tag=2afc8c735218176
    Call-ID: C5570127C1A6A1ABF7ED9DB9AD608CE00xc0a8000a
  ⊿ CSeq: 2 INVITE
        Sequence Number: 2
        Method: INVITE
    Content-Type: application/sdp
    Date: Thu, 12 Apr 2012 15:40:21 GMT
    User-Agent: ENSR3.2.21.22-IS15-RMRG5002-RG900-EP-CPI15-CPO25791
    Content-Length: 236
  ⊿ X-Number-Type: 9055551212;type=off-net
    ⊿ [Expert Info (Note/Undecoded): Unrecognised SIP header (x-number-type)]
        [Unrecognised SIP header (x-number-type)]
        [Severity level: Note]
        [Group: Undecoded]
```

Then, you can see in the body section:

```
⊿ Message Body
  ⊿ Session Description Protocol
        Session Description Protocol Version (v): 0
      ▷ Owner/Creator, Session Id (o): - 819596013 819596013 IN IP4 216.234.64.8
        Session Name (s): ENSResip
      ▷ Connection Information (c): IN IP4 216.234.64.16
      ▷ Time Description, active time (t): 0 0
      ▷ Media Description, name and address (m): audio 54550 RTP/AVP 0 101
      ▷ Media Attribute (a): rtpmap:0 PCMU/8000
      ▷ Media Attribute (a): rtpmap:101 telephone-event/8000
      ▷ Media Attribute (a): fmtp:101 0-11
      ▷ Media Attribute (a): ptime:20
      ▷ Media Attribute (a): setup:active
        Media Attribute (a): sendrecv
```

Here we have **Media Description, name and address (m)**: **audio**; what port is using RTP. So here we are defining in SIP that we'll use RTP to transfer audio traffic between the two devices defined in the message header **From** and **To**.

Once that connection is established, we see that our local phone, this 192 address, is now connected to a different public address; look at that last octet:

55 166.095	0.065078000	192.168.0.10	216.234.64.16	RTP	214	PT=ITU-T G.711 PCMU, SSRC=0x2A173650, Seq=265
57 166.125	0.016285000	192.168.0.10	216.234.64.16	RTP	214	PT=ITU-T G.711 PCMU, SSRC=0x2A173650, Seq=265
58 166.126	0.001239000	192.168.0.10	216.234.64.16	RTP	214	PT=ITU-T G.711 PCMU, SSRC=0x2A173650, Seq=265
59 166.151	0.024678000	216.234.64.16	192.168.0.10	RTP	214	PT=ITU-T G.711 PCMU, SSRC=0x2A173650, Seq=265
62 166.155	0.002953000	192.168.0.10	216.234.64.16	RTP	214	PT=ITU-T G.711 PCMU, SSRC=0x2A173650, Seq=265
63 166.157	0.002844000	216.234.64.16	192.168.0.10	RTP	214	PT=ITU-T G.711 PCMU, SSRC=0x31BE1E0E, Seq=184
64 166.177	0.019945000	216.234.64.16	192.168.0.10	RTP	214	PT=ITU-T G.711 PCMU, SSRC=0x31BE1E0E, Seq=184
65 166.185	0.007168000	192.168.0.10	216.234.64.16	RTP	214	PT=ITU-T G.711 PCMU, SSRC=0x2A173650, Seq=265
66 166.186	0.001204000	192.168.0.10	216.234.64.16	RTP	214	PT=ITU-T G.711 PCMU, SSRC=0x2A173650, Seq=265

We also see that we're using G.711, which is the encoding scheme. There are a number of different encoding schemes out there: some of them are better quality, but with a higher bandwidth; some have less bandwidth with the lesser quality. It's up to you as to which one you want to use, but G.711 is a very common one.

In the RTP information, we can see that we have a **Payload type: ITU-T G.711 PCMU (0)**:

```
⊿ Real-Time Transport Protocol
  ▷ [Stream setup by SDP (frame 54)]
    10.. .... = Version: RFC 1889 Version (2)
    ..0. .... = Padding: False
    ...0 .... = Extension: False
    .... 0000 = Contributing source identifiers count: 0
    1... .... = Marker: True
    Payload type: ITU-T G.711 PCMU (0)
    Sequence number: 26528
    [Extended sequence number: 92064]
    Timestamp: 0
    Synchronization Source identifier: 0x2a173650 (706164304)
    Payload: ff7eff7e7e7e7efefefeffff7e7e7e7e7efffffe7efffeff...
```

ITU is yet another standard body, just like IEEE and IETF. They define certain protocols and encoding schemes such as G.711. Then, we have a whole bunch of data packets between the two devices direct, all with RTP. Then, down at the bottom we have a connection request to terminate the connection, so there's a goodbye:

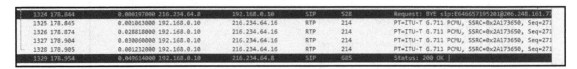

1324 178.844	0.000197000	216.234.64.8	192.168.0.10	SIP	528	Request: BYE sip:E6466571952010206.248.161.77
1325 178.845	0.001063000	192.168.0.10	216.234.64.16	RTP	214	PT=ITU-T G.711 PCMU, SSRC=0x2A173650, Seq=271
1326 178.874	0.028818000	192.168.0.10	216.234.64.16	RTP	214	PT=ITU-T G.711 PCMU, SSRC=0x2A173650, Seq=271
1327 178.904	0.030060000	192.168.0.10	216.234.64.16	RTP	214	PT=ITU-T G.711 PCMU, SSRC=0x2A173650, Seq=271
1328 178.905	0.001232000	192.168.0.10	216.234.64.16	RTP	214	PT=ITU-T G.711 PCMU, SSRC=0x2A173650, Seq=271
1329 178.954	0.049614000	192.168.0.10	216.234.64.8	SIP	685	Status: 200 OK

Then, we have a status `OK`, saying that we terminate the connection. What we can do is take a look at these in the **Telephony** statistics area.

Anything you want to know about your VoIP traffic can be found under this section in Wireshark:

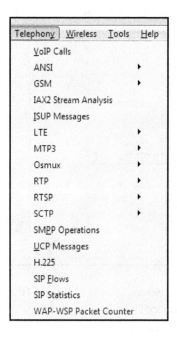

Two of the greatest spots to go look are **VoIP Calls** or **SIP Flows**; it really depends which one you want to go into. But if we go into **VoIP Calls**, we see that here's the one phone call:

If we had a bunch of phone calls happening at the same time in one packet capture, they would all show up here in a nice chart and they would have what times they started and stopped, who initiated the connection, who's it from and to, what protocol were they using, and so on. What we can do is select that and go to **Flow Sequence**, and we can see a great diagram of what ports are on local side on the local phone, what ports are on the remote side on the SIP server, and what ports were used here on the actual device:

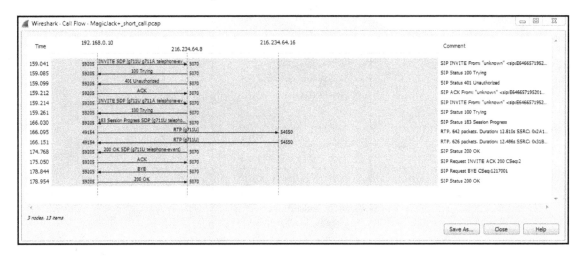

You can see that there are three devices in use. All the SIP traffic (look at the right side in the Comments: you see what protocol's in use) goes to the middleman. Once the connection is established, then we have RTP traffic. The audio traffic in real time goes to and from between these two devices on RTP direct. Then, when we want to terminate the connection, we send a goodbye and then an ok to acknowledge that, and we've terminated the SIP connection. But again, remember it occurred to the control server, the SIP server. This is a really great way of being able to diagnose what's going on in a phone call, and if you have a problem with the SIP connectivity, with some sort of VoIP provider, taking a look at this and telling them what commands are sent and which ones are received is very helpful for the troubleshooting person to diagnose what's going on. They can compare what we see to what they saw and it's very helpful, especially the commands 100 Trying; 200 OK; ACK; and BYE.

In comparison, we can take a look at the **Statistics | Flow Graph**:

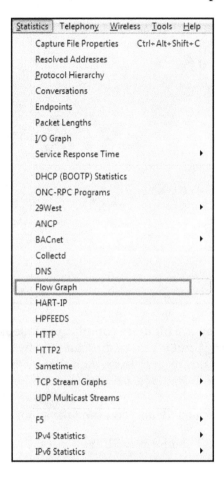

We've looked at this earlier:

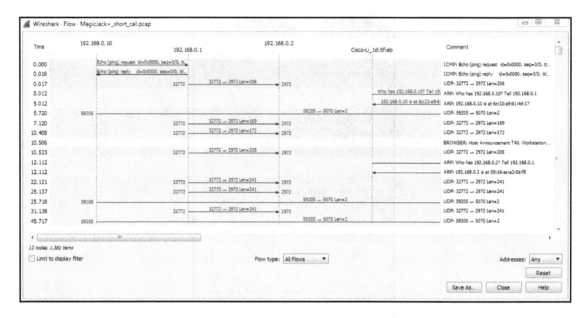

You can see that the flow graph is a bit more complicated to look at. You can still get some of the same information from it, but it's more difficult to view since it's per packet. It's always better to use the other flow graph. Even though we can address this to displayed packets by selecting the option **Displayed packets** from **Show**, it's still for every single packet. So if you're just talking about the basic connectivity or the basic flow, the other flow graph is better. If you need extreme detail, you can take a look at that in here.

You can also take a look at the **Telephony | SIP Statistics**:

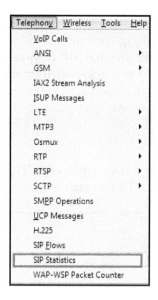

The **SIP Statistics** will give you a count of how many different occurrences have happened of different requests:

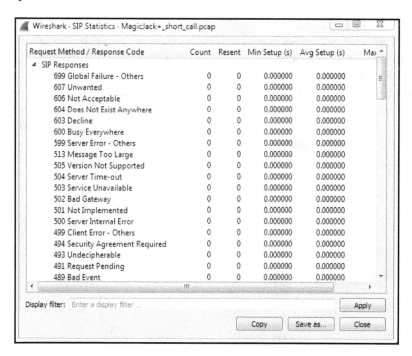

Remember how there are different codes that are in use, just like HTTP and FTP, and all the other ones we've looked at? Same thing here, and it will give us a count of how many times certain things have occurred. So, this was the **SIP Responses**, and then we have **SIP Requests**; and you see that there are two acknowledgments, a goodbye and an unauthorized, and if you have a long packet capture with a bunch of problems, you can take a look at this, and, of course, you can sort that based on your counts and see which ones are the most common. If you see a bunch of 400 or 500 errors, that might be a problem that you need to investigate.

And just like with **SIP Flows**, we can take a look at **RTP**, which is the audio traffic, the actual voice traffic, and look at the **RTP Streams** in addition to the **SIP Streams**:

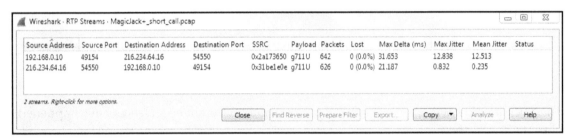

So remember we had that one connection from the phone to the magicJack phone? We see that in the bidirectional setup. So, we have the connection from the phone out to the magicJack, and then from magicJack back to the phone. And these are both RTP; the payload is G.711, like we talked about before. It also gives us some good information about the jitter. The jitter is a very useful statistic to look at. The lower the jitter, the better things are for voice traffic. Any sort of packet loss that you have (which is also a column shown earlier); any sort of jitter that you have; or any latency: these are all major problems for voice traffic, be it any sort of real-time traffic: video traffic, voice traffic, and so on.

Jitter is the difference in latency between different packets, and you have massive swings from one packet to another. You can have a lot of audio quality problems as packets are arriving at different times and get jumbled up in the software. You might have some packet loss as well in that sort of situation, so then you're losing certain words, and that's when you start getting really weird crackling problems; you get dropped words; you might have one side that's completely silent and the other side can hear. You can have all sorts of strange problems when you have a lot of jitter or packet loss. In this section, we looked at some VoIP traffic using SIP and RTP and how to filter on those, as well as how to look at some statistics and some flow graphs on them.

VoIP playback

In this section, we'll reconstruct and play back VoIP calls and listen to quality issues.

What we'll do is use the same magicJack call that was used in the previous section.

One of the really great features of Wireshark, in addition to all of its many filters and statistics and graphs that it can create, is that it has the ability to play back voice traffic. Some people might find this kind of creepy if you're an end user, that you can listen to someone's phone call, but it is data. Just like we can read all of these commands back and forth, we can read the username and password if it's unencrypted; if the SIP traffic is also unencrypted then we can listen to the phone call, too, just because it's standard data; it just happens to be voice traffic. We can do that in Wireshark very easily.

In order to do that, we opened up our call. We simply go to **Telephony** | **VoIP Calls**, and select the call:

After selecting the call, click on **Play Streams**. And when you click on that, you'll see a histogram of exactly what's going on in the voice call:

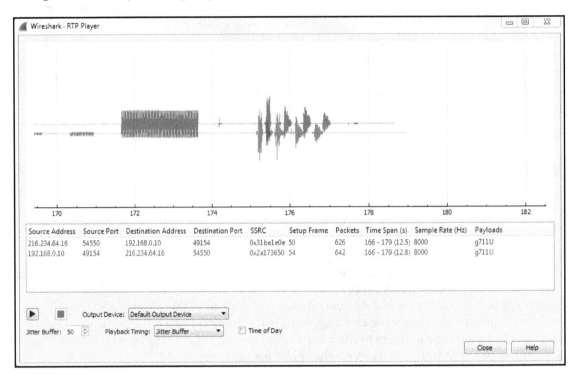

This alone may be able to tell you something about what's going on, once you get used to what certain things look like. If it's a phone that's ringing, or actual words and discussions, you might be able to pick that out just by looking at the histogram shown in the preceding screenshot. And then, of course, you can select one side or the other to bring it to your attention, so you can see what might be occurring on one of these channels or the other, whether it's the source to destination or the destination to the source from one phone to another. What we can do is, leave everything as default, and go ahead and click on the play button.

You can hear the phone ringing, after which it says Test, 1, 2, 3.

You'll see that it played back the audio of the phone call. It played back both directions of that phone call, so we got to hear both the sender side and the receiver side. So, what we had first in the top was one side of the conversation, and then below it is the other side of the conversation. So, that's very useful, to be able to actually hear both sides of the call at the same time. That way, if one user is complaining about something and the other one isn't, you can actually listen to that as if you had both telephones up to both of your ears at the same time.

What you can also do is adjust the jitter in the player, and listen to what that might sound like and what kind of problems might occur.

Now, you will notice that the preceding graph was real-time, and as the audio plays, you will see the playback line go across. This is a relatively clean capture. If there are any problems in the capture, they will show up on this graph, and you'll see them begin to occur as you change things such as the **Jitter Buffer**. So what we'll do is we'll drop that down to something very low, let's say 5:

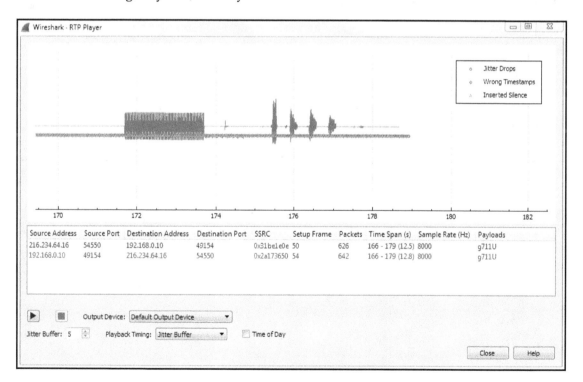

You can already see that we have some silence that was created, as well as some additional errors. If we zoom in, we have some jitter drops that have also been created, and we've actually manufactured some problems in our connection:

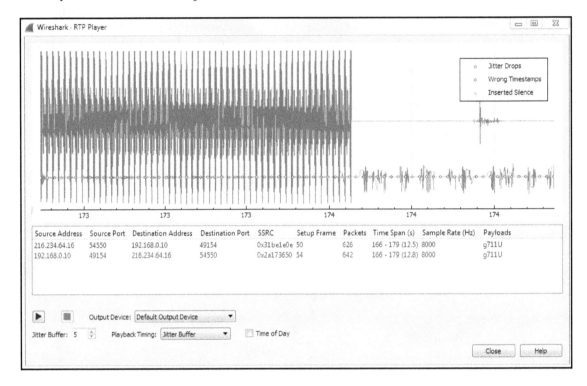

And, if you zoom out far enough, you'll see there's that initial ringtone that occurred, which was off screen before:

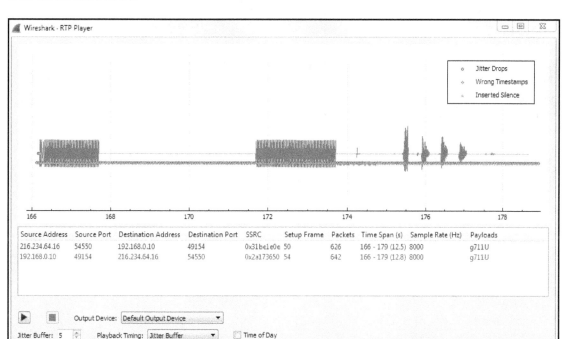

Now that we've inserted all of these problems in our packet capture, let's go ahead and play it.

You can hear the phone ringing and it says `Test 1, 2, 3`.

This time you should be able to hear that difference. In the beginning of the phone call, we had a bit of a crackly sound. As the phone was ringing we also had some kind of dropped packets there; you could hear it kind of was crackly and didn't sound that great. And the voice in the beginning there also had some words that were kind of clipped and missing. Additionally, one side is completely silent. If you remember looking at the histogram, there was a duplicate of each word in the test 1, 2, 3. One side is now completely silent; it's dropping all of that and is now completely missed because there are too many problems in the jitter. So this is a great tool to use to be able to recreate different problems and be able to listen to what they actually sound like in real life.

Summary

So in this chapter, you've learned about some email analysis using POP and SMTP. We also looked at 802.11, which is wireless, what certain things look like in a packet capture, and the fact that you'll need to do additional troubleshooting outside of what you can do in Wireshark in order to properly diagnose wireless. We also looked at VoIP analysis using SIP and RTP, creating that connection and then transmitting that audio data directly from one device to another. Then, we also played that back using the built-in tools in Wireshark, and manipulated some of the settings in that for jitter, in order to recreate problems and be able to listen to what that sounds like.

Next is Chapter 10, *Command-Line Tools*, where we'll use some command-line tools to extend Wireshark, and talk about some of the enhancements that you can add to it.

Command-Line Tools **10**

In this chapter, we'll take a look at the following topics:

- Running Wireshark from a command line
- Running tshark
- Running tcpdump
- Running dumpcap

Running Wireshark from a command line

In this section, we'll take a look at how to run Wireshark from a command line and explore some of the command-line options and how you might use them. The first thing I want to do is open up a Command Prompt, and then we'll browse where Wireshark is. Unless you have Wireshark in your system variable, you won't be able to simply type `wireshark` and have that function.

So what we'll have to do is go to its location. In my system, it's back in program files and in the `wireshark` directory. And if we type `dir`, we'll see `Wireshark.exe`, as well as some of the other tools that we'll talk about later, such as tshark:

```
🔲 Command Prompt                                                    ─  ▢  ⛬

04/24/2018    11:09 PM                    821  README.windows.txt
04/24/2018    11:23 PM                327,848  reordercap.exe
04/24/2018    11:08 PM                270,998  services
04/24/2018    11:08 PM                    333  smi_modules
04/27/2018    03:52 PM    <DIR>               snmp
04/24/2018    11:23 PM                351,912  text2pcap.exe
04/24/2018    11:08 PM                 13,032  text2pcap.html
04/27/2018    03:52 PM    <DIR>               tpncp
04/27/2018    03:52 PM    <DIR>               translations
04/24/2018    11:23 PM                583,848  tshark.exe
04/24/2018    11:08 PM                104,029  tshark.html
04/24/2018    11:23 PM                431,392  uninstall.exe
04/24/2018    11:19 PM              3,661,788  user-guide.chm
02/14/2018    03:04 AM             15,328,616  vcredist_x64.exe
04/27/2018    03:52 PM    <DIR>               wimaxasncp
04/03/2018    02:10 AM                915,128  WinPcap_4_1_3.exe
04/24/2018    11:23 PM              1,868,456  WinSparkle.dll
04/24/2018    11:08 PM                 21,293  wireshark-filter.html
04/24/2018    11:23 PM              8,125,608  Wireshark.exe
04/24/2018    11:08 PM                198,255  wireshark.html
04/24/2018    11:08 PM                 10,720  wka
04/24/2018    11:08 PM                 40,370  ws.css
04/24/2018    11:23 PM                147,112  zlib1.dll
              93 File(s)      158,803,248 bytes
              21 Dir(s)   156,652,003,328 bytes free

C:\Program Files\Wireshark>
```

What we'll do is run `Wireshark.exe`; then, if you press *Enter*, it will open up Wireshark, just like if you were to click on the icon. If you type `Wireshark.exe -h`, it will provide the output of all the variables and arguments that Wireshark has available to it:

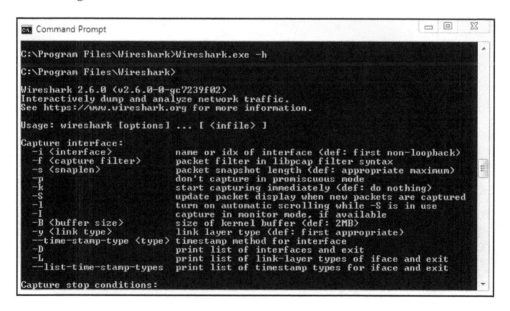

If we scroll up, we'll see the version of Wireshark we're running and a dump of all the variables and arguments that we can use:

You'll see that they're broken up into the following different categories:

- `Capture interface`
- `Capture stop conditions`
- `Capture output`
- `Processing`
- `User interface`
- `Output`
- `Miscellaneous`

One of the first things you'll most likely want to do is set up Wireshark to run a capture with your default interface, your standard local area connection wired interface. You should see that you would do it with `Wireshark -i` and then the interface. It says the interface name or `idx`. Now, how do you determine what the interface name is or what `idx` is - what the index of what that interface is? This, you do with `-D`, which prints the list of interfaces and exits. Now, we will run `Wireshark.exe -D`.

Note that the capitalization does matter.

We see that in the following screenshot, we have index numbers. That's idx: 1, 2, 3, 4, and 5. After the unique ID of that interface, we have the name, such as `Local Area Connection`, `VirtualBox Host-Only Network`, and other stuff that's on my system:

```
Command Prompt

C:\Program Files\Wireshark>Wireshark.exe -D

C:\Program Files\Wireshark>

1. \Device\NPF_{F2E09F0C-9693-4C5B-BDA9-81F8B4FC2D6C} (VirtualBox Host-Only Netw
ork)
2. \Device\NPF_{A8B1E1DA-0270-446C-B13A-FA4AAAB4CD41} (Local Area Connection)
3. \\.\USBPcap1 (USBPcap1)
4. \\.\USBPcap2 (USBPcap2)
5. \\.\USBPcap3 (USBPcap3)
Vir
```

We can then execute `Wireshark.exe -i` and then either the name or the idx number. We will use the index number because it's nice and short and it prevents any additional typos. We'll type `Wireshark.exe -i 1`. Simply press *Enter*, and you'll see that it doesn't really do much. This is because we've opened up Wireshark, and by default, Wireshark opens up to the main home screen. The main home screen doesn't actually have to do anything yet at this point and so, it doesn't. You're not starting a capture; you're not doing anything. You just open it up. Now, you can see that it does have the first interface kind of selected, but it's not that useful. So, let's close this, and we'll tell Wireshark to start capturing the second it opens up, using the interface that we've defined. We can do that with `-k`, which says `start capturing immediately (def: do nothing)`. It just opens up the Wireshark interface. We'll use `-k`. Type `Wireshark.exe -i 1 -k`, and it will immediately start capturing on that interface.

Another useful feature is to run Wireshark and have it automatically create files after a certain filesize or a certain duration of time, or after a certain number of bytes, or anything like that. Most commonly, people run Wireshark in a ring buffer, where you create a file and then it automatically switches to a new file every hour, every 30 minutes, or every 10 MB; whatever it might be that you have decided is the best way to approach this issue and to approach this system that you're capturing, based on the volume of data that you're expecting. What we can do is execute Wireshark using the flag to allow us to automatically create files. This is very useful if you want this to run overnight and not crash your system by having a 20 GB capture file in the morning: that'll not be very useful and it'd be very difficult to work with, if it is even possible at all, but it depends on your system. We'll run Wireshark with the primary interface that we've already selected, `-i 1 -k`, so it initiates the packet capture right away. Then, we'll define the fact that we're using a ring buffer. So, we'll use `-b` for ring buffer and the kind of ring buffer that we want. Do we want filesize? Do we want duration? What might it be? You'll see that ring buffer has a number of options such as duration, filesize, and the number of files that you want to replace them; hence the term **ring**, where they replace themselves in the circle. What we'll do is select `-b` for ring buffer, and we'll use `filesize 10000`.

If you do `filesize 10000`, generally the filesize is based on KB. We're saying that approximately 10 MB makes a new file, so every 10 MB we'll make a new capture file. We then have to define the output file. So, we're asking for a ring buffer, but we need to tell it where to put it. We'll say `-w` for write to (we'll write to an output file), and it'll be located in `C:\Users\sayalit\capture.pcapng`:

```
Wireshark.exe -i 1 -k -b filesize 10000 -w C:\Users\sayalit\capture.pcapng
```

If we run the preceding command, that will have Wireshark run and begin capturing all the packets. Once it hits 10 MB, it will then refresh the screen and start capturing again, and you'll see it just kind of flicker once every time it hits 10 MB and keeps coming up with a new screen as it captures a new cap file. If you go into the directory that was specified in the `-w` argument, you'll see all the files that were created every time it hits 10 MB. This was just an example of a few arguments and flags you can use when running Wireshark from the command line. As you saw, there are many options, and you can create very elaborate custom ones if you wish. Up next, we'll look at tshark, which is a command-line-only application that comes with Wireshark when it's installed.

Running tshark

In this section, we'll take a look at how to run the terminal version of Wireshark, so that it only has a command-line interface instead of opening up the GUI.

In order to run tshark, you have to open up the command window, and once it's up, we have to browse to where Wireshark is installed because as I've explained, unless you have it in your system path, it'll not be available. So we'll browse again to where Wireshark lives, and we'll do a directory listing. We'll see that we have `tshark.exe`. This is installed by default with Wireshark. In order to run tshark, all you have to do is, of course, run `tshark.exe`. If you do so, it automatically begins capturing on your default interface:

You'll notice that it shows the packets that it's capturing directly to the command-line interface, directly to `stdout`. It does so because it does not have a graphical interface; there's nothing for it to display except for the screen that it's currently using, which is the command interface. You'll see that the output provides a similar display as you would see in Wireshark in the GUI. We have the packet number, the time since the packet capture started, the time difference between the last two packets, the source IP and the source port, the destination IP, the destination port, and so on. If you take a look at `tshark.exe -h`, just like `Wireshark.exe`, it'll look very similar. If we scroll up, we have the same arguments that we can use:

```
Command Prompt                                                    ⸺  ▣  ☒

C:\Program Files\Wireshark>tshark.exe -h
TShark (Wireshark) 2.6.0 (v2.6.0-0-gc7239f02)
Dump and analyze network traffic.
See https://www.wireshark.org for more information.

Usage: tshark [options] ...

Capture interface:
  -i <interface>           name or idx of interface (def: first non-loopback)
  -f <capture filter>      packet filter in libpcap filter syntax
  -s <snaplen>             packet snapshot length (def: appropriate maximum)
  -p                       don't capture in promiscuous mode
  -I                       capture in monitor mode, if available
  -B <buffer size>         size of kernel buffer (def: 2MB)
  -y <link type>           link layer type (def: first appropriate)
  --time-stamp-type <type> timestamp method for interface
  -D                       print list of interfaces and exit
  -L                       print list of link-layer types of iface and exit
  --list-time-stamp-types  print list of timestamp types for iface and exit

Capture stop conditions:
  -c <packet count>        stop after n packets (def: infinite)
  -a <autostop cond.> ...  duration:NUM - stop after NUM seconds
                           filesize:NUM - stop this file after NUM KB
                               files:NUM - stop after NUM files
Capture output:
  -b <ringbuffer opt.> ...  duration:NUM - switch to next file after NUM secs
                           interval:NUM - create time intervals of NUM secs
                           filesize:NUM - switch to next file after NUM KB
                               files:NUM - ringbuffer: replace after NUM files
RPCAP options:
  -A <user>:<password>     use RPCAP password authentication
Input file:
  -r <infile>              set the filename to read from (- to read from stdin)
```

We have `-D` to display the list of interfaces again and `-i` for the interface that we want to use. We don't have to define `-k` so that it automatically starts capturing because tshark doesn't have a GUI for us to do anything in, so it automatically starts capturing anyway. We can set the ring buffer as before; we can define output files; we can do all sorts of things.

For an example, we can type `tshark -D` just to display all of our interfaces again. To confirm that we want to use interface number 1, we'll type `tshark.exe -i 1`, which ensures the use of the first interface. Then, we can define an output file as well, so we'll write this out to `C:\Users\sayalit\test.pcapng`, and now it begins capturing:

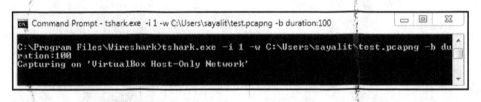

You'll see that it's showing us how many packets it's captured as it runs. In order to stop the capture, simply press *Ctrl + C*, and it will stop whatever it's doing. Just like with the Wireshark example, we can enhance this line that we've already created and we can define a ring buffer, for example, and say the duration is every `100` seconds. Then, it will do exactly that; every `100` seconds, it'll create a new file:

```
Command Prompt - tshark.exe  -i 1 -w C:\Users\sayalit\test.pcapng -b duration:100
C:\Program Files\Wireshark>tshark.exe -i 1 -w C:\Users\sayalit\test.pcapng -b du
ration:100
Capturing on 'VirtualBox Host-Only Network'
```

Tshark is very useful for things that you want to script. If you want to write a batch script or a bash script that will do a capture, and you want to ensure that it uses all the functionality that Wireshark is capable of, and it saves to the `pcapng` format, and you want to make sure that it does everything just like you would in Wireshark itself, using tshark is a great idea, as well as on systems that have low resources. If you're running this on a Command Prompt, need to do a packet capture on some old system like a Windows 2000 system, or something like that that's barely scraping by, you can run `tshark`, and it will eliminate a lot of the overheads that you have with running Wireshark, especially if the interface is automatically updating and scrolling with the packets. This gets rid of all of that, and it just gets the data that we need. In our next section, we'll take a look at `tcpdump`, which is available on almost every Unix or Linux system out there.

Running tcpdump

In this section, we'll take a look at how to run tcpdump on a Linux system to capture traffic.

If you have a Linux- or a Unix-based system (BSD; whatever it might be) that does not have Wireshark installed and you do not have the option of installing Wireshark, or if you have a system where you don't really want to spend the time to install Wireshark and you just want to do a quick capture, you can do so on almost all of them with tcpdump. This is a very common utility that's installed on almost every single NIC-based system out there.

What we have is a newer version of Ubuntu, and I've opened up the Terminal window, and all you have to do is run tcpdump. It's within the system variable path, so you don't have to go browse for it like we had to for the others on Windows, and I'll run it with --help. We can see that tcpdump has displayed its help contents and it shows us what arguments are available for it to receive:

```
andrew@ubuntu:~$ tcpdump --help
tcpdump version 4.7.4
libpcap version 1.7.4
OpenSSL 1.0.2g  1 Mar 2016
Usage: tcpdump [-aAbdDefhHIJKlLnNOpqRStuUvxX#] [ -B size ] [ -c count ]
               [ -C file_size ] [ -E algo:secret ] [ -F file ] [ -G seconds ]
               [ -i interface ] [ -j tstamptype ] [ -M secret ] [ --number ]
               [ -Q in|out|inout ]
               [ -r file ] [ -s snaplen ] [ --time-stamp-precision precision ]
               [ --immediate-mode ] [ -T type ] [ --version ] [ -V file ]
               [ -w file ] [ -W filecount ] [ -y datalinktype ] [ -z command ]
               [ -Z user ] [ expression ]
```

If you want to learn more about tcpdump within the Terminal window, if you're not familiar with Linux, you can do man tcpdump, and it will provide you the manual on how to use tcpdump. It is a nice, long document describing all the different arguments, what they do, and how to use it, with some examples and such. To get out of there, you just need to press *Q*.

Now, you'll notice in the `Usage` line that the syntax for `tcpdump` and arguments and the flags that it has available are different from tshark and Wireshark, so we can't run the exact same commands. They're similar, but they're not exactly the same. For example, there's no `-D` for us to take a look at our interfaces. If you want to know the interfaces on your Linux system, usually you can run something like `ifconfig` or `iwconfig`, depending on what you're looking for, and it will output the interfaces that you have available, the IP addresses associated with them, and a bunch of other statistics:

```
andrew@ubuntu:~$ man tcpdump
andrew@ubuntu:~$ ifconfig
ens33     Link encap:Ethernet  HWaddr 00:0c:29:5a:28:8e
          inet addr:192.168.139.131  Bcast:192.168.139.255  Mask:255.255.255.0
          inet6 addr: fe80::ffe8:9443:3f8f:42d0/64 Scope:Link
          UP BROADCAST RUNNING MULTICAST  MTU:1500  Metric:1
          RX packets:167879 errors:0 dropped:0 overruns:0 frame:0
          TX packets:75044 errors:0 dropped:0 overruns:0 carrier:0
          collisions:0 txqueuelen:1000
          RX bytes:220455267 (220.4 MB)  TX bytes:4541715 (4.5 MB)

lo        Link encap:Local Loopback
          inet addr:127.0.0.1  Mask:255.0.0.0
          inet6 addr: ::1/128 Scope:Host
          UP LOOPBACK RUNNING  MTU:65536  Metric:1
          RX packets:275 errors:0 dropped:0 overruns:0 frame:0
          TX packets:275 errors:0 dropped:0 overruns:0 carrier:0
          collisions:0 txqueuelen:1
          RX bytes:22103 (22.1 KB)  TX bytes:22103 (22.1 KB)
```

Now, this is a virtual machine that's running right now, which is why it shows `ens33` as the interface name. Quite often, this is `eth0` or something like that. `lo` is our loopback on the system. Just like every other system you've seen in Windows and such, there's usually a loopback. This is a loopback on this system. If we look back to our information, we can run `tcpdump`, and we can also define an interface with the `-i` so that's where that interface comes in. We're going to do `tcpdump -i ens33`. Now, by default, `tcpdump` without the `-i` command will try to run on the default interface that you have, the `ens33` interface in this example, and we can take a look at that in the following screenshot. When I try to do that, you'll see that I have an error. It says that `You don't have permission to capture on that device`:

```
andrew@ubuntu:~$ ifconfig
ens33     Link encap:Ethernet  HWaddr 00:0c:29:5a:28:8e
          inet addr:192.168.139.131  Bcast:192.168.139.255  Mask:255.255.255.0
          inet6 addr: fe80::ffe8:9443:3f8f:42d0/64 Scope:Link
          UP BROADCAST RUNNING MULTICAST  MTU:1500  Metric:1
          RX packets:167879 errors:0 dropped:0 overruns:0 frame:0
          TX packets:75044 errors:0 dropped:0 overruns:0 carrier:0
          collisions:0 txqueuelen:1000
          RX bytes:220455267 (220.4 MB)  TX bytes:4541715 (4.5 MB)

lo        Link encap:Local Loopback
          inet addr:127.0.0.1  Mask:255.0.0.0
          inet6 addr: ::1/128 Scope:Host
          UP LOOPBACK RUNNING  MTU:65536  Metric:1
          RX packets:275 errors:0 dropped:0 overruns:0 frame:0
          TX packets:275 errors:0 dropped:0 overruns:0 carrier:0
          collisions:0 txqueuelen:1
          RX bytes:22103 (22.1 KB)  TX bytes:22103 (22.1 KB)

andrew@ubuntu:~$
andrew@ubuntu:~$ tcpdump
tcpdump: ens33: You don't have permission to capture on that device
(socket: Operation not permitted)
andrew@ubuntu:~$
```

Depending on your system, you'll probably have to run `sudo` and in order to do this, enter the password for your user which will elevate the privileges for your user so that you can run this command. You can see that it says `tcpdump` is now running. It would show us right now any packets that are coming into and out of the system if they were, but because this is the virtualized system, `tcpdump` is not functioning correctly on it right now:

```
          UP BROADCAST RUNNING MULTICAST  MTU:1500  Metric:1
          RX packets:167879 errors:0 dropped:0 overruns:0 frame:0
          TX packets:75044 errors:0 dropped:0 overruns:0 carrier:0
          collisions:0 txqueuelen:1000
          RX bytes:220455267 (220.4 MB)  TX bytes:4541715 (4.5 MB)

lo        Link encap:Local Loopback
          inet addr:127.0.0.1  Mask:255.0.0.0
          inet6 addr: ::1/128 Scope:Host
          UP LOOPBACK RUNNING  MTU:65536  Metric:1
          RX packets:275 errors:0 dropped:0 overruns:0 frame:0
          TX packets:275 errors:0 dropped:0 overruns:0 carrier:0
          collisions:0 txqueuelen:1
          RX bytes:22103 (22.1 KB)  TX bytes:22103 (22.1 KB)

andrew@ubuntu:~$
andrew@ubuntu:~$ tcpdump
tcpdump: ens33: You don't have permission to capture on that device
(socket: Operation not permitted)
andrew@ubuntu:~$ sudo tcpdump
[sudo] password for andrew:
tcpdump: verbose output suppressed, use -v or -vv for full protocol decode
listening on ens33, link-type EN10MB (Ethernet), capture size 262144 bytes
```

Press *Ctrl + C* to cancel it, and you see that there are 0 packets. What we'll do instead is define the loopback interface for the example, using `tcpdump -i lo` for loopback, as we saw up here under the `ifconfig` command. If I do that now, it's listening; but, we don't have any traffic yet that is going to the loopback:

```
            inet addr:127.0.0.1  Mask:255.0.0.0
            inet6 addr: ::1/128 Scope:Host
            UP LOOPBACK RUNNING  MTU:65536  Metric:1
            RX packets:275 errors:0 dropped:0 overruns:0 frame:0
            TX packets:275 errors:0 dropped:0 overruns:0 carrier:0
            collisions:0 txqueuelen:1
            RX bytes:22103 (22.1 KB)  TX bytes:22103 (22.1 KB)

andrew@ubuntu:~$
andrew@ubuntu:~$ tcpdump
tcpdump: ens33: You don't have permission to capture on that device
(socket: Operation not permitted)
andrew@ubuntu:~$ sudo tcpdump
[sudo] password for andrew:
tcpdump: verbose output suppressed, use -v or -vv for full protocol decode
listening on ens33, link-type EN10MB (Ethernet), capture size 262144 bytes
^C
0 packets captured
0 packets received by filter
0 packets dropped by kernel
andrew@ubuntu:~$ sudo tcpdump -i lo
tcpdump: verbose output suppressed, use -v or -vv for full protocol decode
listening on lo, link-type EN10MB (Ethernet), capture size 262144 bytes
```

Now, we'll open up a new Terminal, and I'll generate some traffic. I'll simply `ping` my `localhost`. `localhost` as an alias name to `127.0.0.1`, which is my loopback address. As I ping loopback, you'll now see that popping up in the `tcpdump` window:

```
andrew@ubuntu:~$ sudo tcpdump -i lo
tcpdump: verbose output suppressed, use -v or -vv for full protocol decode
listening on lo, link-type EN10MB (Ethernet), capture size 262144 bytes
18:59:45.040066 IP localhost > localhost: ICMP echo request, id 19418, seq 1, le
ngth 64
18:59:45.040078 IP localhost > localhost: ICMP echo reply, id 19418, seq 1, leng
th 64
18:59:46.039267 IP localhost > localhost: ICMP echo request, id 19418, seq 2, le
ngth 64
18:59:46.039278 IP localhost > localhost: ICMP echo re
th 64                                                    andrew@ubuntu: ~
18:59:47.039418 IP localhost > localhost: ICMP echo r andrew@ubuntu:~$ ping localhost
ngth 64                                                PING localhost (127.0.0.1) 56(84) bytes of data.
18:59:47.039429 IP localhost > localhost: ICMP echo r 64 bytes from localhost (127.0.0.1): icmp_seq=1 ttl=64 time=0.084 ms
th 64                                                  64 bytes from localhost (127.0.0.1): icmp_seq=2 ttl=64 time=0.051 ms
18:59:48.039324 IP localhost > localhost: ICMP echo r 64 bytes from localhost (127.0.0.1): icmp_seq=3 ttl=64 time=0.046 ms
ngth 64                                                64 bytes from localhost (127.0.0.1): icmp_seq=4 ttl=64 time=0.049 ms
18:59:48.039334 IP localhost > localhost: ICMP echo re
th 64
```

You see that, similar to tshark and Wireshark, we have the time that the packet occurred, from and to address, what kind of protocol is it, what's in the packet, and the details in there. We'll cancel that. *Ctrl + C* closes both of those.

Now if I were to do a listing, this is like `dir` in Windows: `ls`; I don't have any files here besides the test one that I did earlier:

```
19:00:08.039088 IP localhost > localhost: ICMP echo request, id 19418, seq 24, l
ength 64
19:00:08.039102 IP localhost > localhost: ICMP echo reply, id 19418, seq 24, len
gth 64
19:00:09.039182 IP localhost > localhost: ICMP echo request, id 19418, seq 25, l
ength 64
19:00:09.039193 IP localhost > localhost: ICMP echo reply, id 19418, seq 25, len
gth 64
19:00:10.039226 IP localhost > localhost: ICMP echo request, id 19418, seq 26, l
ength 64
19:00:10.039237 IP localhost > localhost: ICMP echo reply, id 19418, seq 26, len
gth 64
19:00:11.039182 IP localhost > localhost: ICMP echo request, id 19418, seq 27, l
ength 64
19:00:11.039194 IP localhost > localhost: ICMP echo reply, id 19418, seq 27, len
gth 64
^C
54 packets captured
108 packets received by filter
0 packets dropped by kernel
andrew@ubuntu:~$ ls
Desktop    Downloads          Music     Public     test.pcap
Documents  examples.desktop   Pictures  Templates  Videos
andrew@ubuntu:~$
```

By running that, it outputs to the `stdout` to the Terminal, but it doesn't actually save the file. In order to do so, we would have to define the file, just like in tshark. We'll remove my previous test files by typing the command `rm test.pcap`; you'll be able to see that it's now gone. We'll run `tcpdump` again, and this time, we'll define an output file for it to save to. As with tshark, luckily, we use `-w` for write. So, we're writing to a file `test.pcap`. After typing the command `sudo tcpdump -i lo -w test.pcap` and starting to generate some traffic, instead of outputting to the `stdout`, it's outputting to the file, and I can stop my traffic. I can also stop my capture, and it says it's saved `64` packets:

```
19:00:10.039237 IP localhost > localhost: ICMP echo reply, id 19418, seq 26, len
gth 64
19:00:11.039182 IP localhost > localhost: ICMP echo request, id 19418, seq 27, l
ength 64
19:00:11.039194 IP localhost > localhost: ICMP echo reply, id 19418, seq 27, len
gth 64
^C
54 packets captured
108 packets received by filter
0 packets dropped by kernel
andrew@ubuntu:~$ ls
Desktop    Downloads          Music     Public     test.pcap
Documents  examples.desktop   Pictures  Templates  Videos
andrew@ubuntu:~$ rm test.pcap
rm: remove write-protected regular file 'test.pcap'? yes
andrew@ubuntu:~$ ls
Desktop    Downloads          Music     Public     Videos
Documents  examples.desktop   Pictures  Templates
andrew@ubuntu:~$ sudo tcpdump -i lo -w test.pcap
tcpdump: listening on lo, link-type EN10MB (Ethernet), capture size 262144 bytes
^C32 packets captured
64 packets received by filter
0 packets dropped by kernel
andrew@ubuntu:~$ ls
```

If I take a look at my listing, I do have `test.pcap`, and if I look at the size, I'll see that it does indeed have some bytes in there. There's also an option with `W`, and I wanted to point this out because I've been showing examples of ring buffers. This is a way of doing a ring buffer where it will automatically overwrite so many files. However, it will save many files you define and then, when you get to the maximum number, it will start overwriting the last oldest files. In the next section, we'll take a look at `dumpcap`, which is another option for `tshark` or `tcpdump`.

Running dumpcap

In this section, we'll take a look at how to run `dumpcap`, which is another alternative to `tshark` and `tcpdump`.

Once again, we'll have to go to `dumpcap`. In this example, it is installed with Wireshark on the system, and if we do a directory listing, you'll see that `dumpcap` is indeed listed. Tshark is actually based on `dumpcap`, and so we can type `dumpcap.exe --help` or `-h`. If we take a look at the output, it looks very similar to tshark and Wireshark:

Depending on the system that you're using though, it may only have dumpcap available for one reason or another, or tshark may be using too much memory if it's a really small, embedded IoT system, or something like that. You could potentially use dumpcap to have an even lighter utility in order to capture traffic-or maybe you just like using this better. If we look at the arguments that are available, they are just like in tshark. We have -i, -D, and -w for output. They're all very much the same. We can illustrate that by running dumpcap, and we'll do -D to display our interfaces again. We can type dumpcap -i 1, and we'll output to a file again. We'll type C:\Users\sayalit\dump.pcap. If we start doing that, it'll begin capturing packets:

```
C:\Program Files\Wireshark>dumpcap.exe -D
1. \Device\NPF_{F2E09F0C-9693-4C5B-BDA9-81F8B4FC2D6C} (VirtualBox Host-Only Netw
ork)
2. \Device\NPF_{A8B1E1DA-0270-446C-B13A-FA4AAAB4CD41} (Local Area Connection)

C:\Program Files\Wireshark>dumpcap.exe -i 1 -w C:\Users\sayalit\dump.pcap
Capturing on 'VirtualBox Host-Only Network'
File: C:\Users\sayalit\dump.pcap
Packets captured: 0
Packets received/dropped on interface 'VirtualBox Host-Only Network': 0/0 (pcap:
0/dumpcap:0/flushed:0/ps_ifdrop:0) (0.0%)
```

We can also expand this and define a buffer to use as well, just like the other examples, and we can do -b and say a duration of 60 seconds. This way, every minute, it will create a new file. We can also illustrate this by reducing the number of seconds. Let's change that to 5. Every 5 seconds, it'll create a new file for us. The filename is changing every 5 seconds. This is an example of using that buffer. Note that in all of these utilities, you can define filters that can restrict the capture, so you can apply capture filters to these. I would not recommend doing that unless you really have to. It's much better, if possible, to capture everything and then filter only what you need with the display filters in Wireshark. By applying capture filters, you can potentially miss some of the packets that might be useful. Maybe you're only capturing TCP traffic, but then a whole big blast of ARP or ICMP traffic ends up causing a problem. You would miss that because you're not capturing that traffic. If possible, you should just capture what you can with the defaults, use just a basic interface selection and maybe a ring buffer, and take the files and parse them out. Also, observe them in Wireshark. Just apply display filters and carry out the necessary actions such as graphing it.

Summary

In this chapter, we went over several command-line options for Wireshark. We discussed running Wireshark from the command line and some of the arguments that are available with it. We also discussed running tshark, which is the command-line version of Wireshark, running tcpdump, which is a generic dumping utility that's available on many Linux and Unix systems, as well as running dumpcap, which is another alternative for tshark and Wireshark.

In Chapter 11, *A Troubleshooting Scenario*, we'll dive into an issue with the user trying to connect to an FTP server.

11
A Troubleshooting Scenario

In this chapter, we'll take a look at troubleshooting a specific issue within Wireshark. We will do so by performing the following steps:

- Expand Wireshark with some additional plugins and dissectors
- Determine where to begin our packet capture about the troubleshooting issues that we'll look at
- Capture the actual traffic
- Diagnose the traffic

Wireshark plugins

In this section, we'll take a look at Wireshark plugins that are available and see how to develop them.

Now, plugins in Wireshark are dissectors, and dissectors are different ways for Wireshark to analyze and take apart different protocols. If, for some reason, Wireshark is unable to interpret the data you are capturing, you can look for additional dissectors that are out there, or write your own, in order to accomplish what you wish to accomplish.

The Wireshark wiki has a page here on dissectors (`https://wiki.wireshark.org/Lua/Dissectors`):

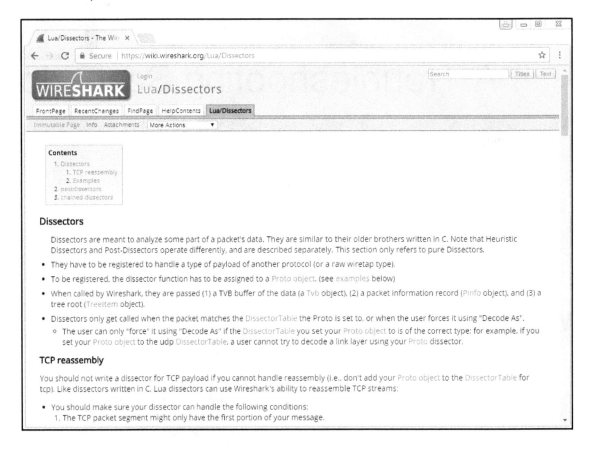

It explains how, from a programmatic standpoint, you would handle designing and creating a dissector. You can see on the page that it goes through many of the details in order to do so. It has some great information on how you would handle a dissector, and what it would do and how you'd go about creating it, but it doesn't actually tell you the individual API data. You can go to the Wireshark developer's guide in order to do so (`https://www.wireshark.org/docs/wsdg_html_chunked/`):

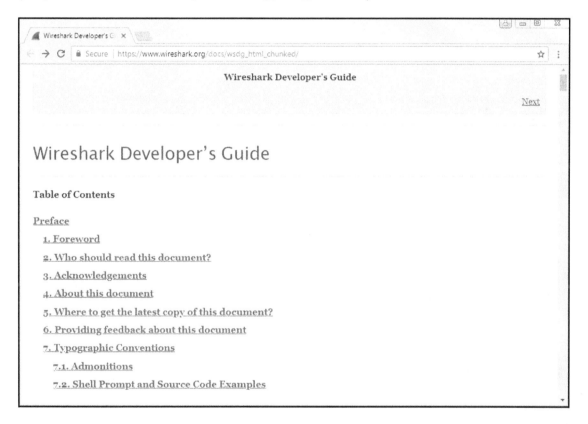

In the Wireshark developer's guide, scroll down to the correct section. You'll see under **Wireshark Development** we have **Packet dissection**. Let's click on it and see what we get:

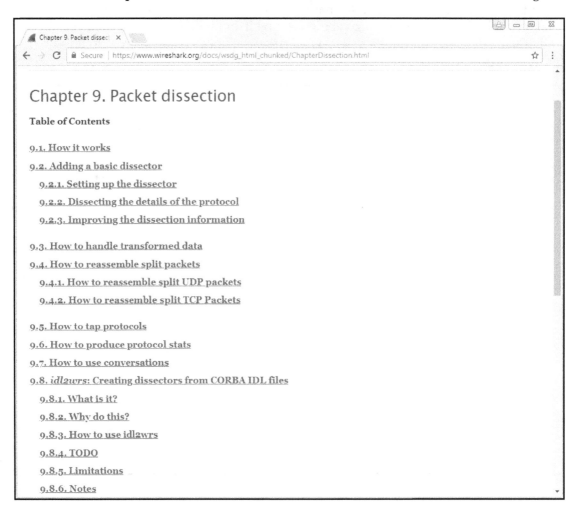

You can see it gives you an entire write-up on how dissection works within Wireshark and how you can expand upon it with Lua in order to create your own dissectors. Additionally, if you go back and go down to Section 11, you can see that there is the entire **Wireshark Lua API Reference Manual**:

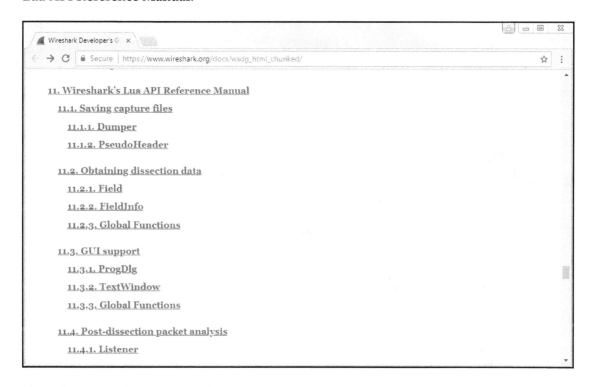

If you happen to know Lua and you're programming your dissector, you know how to reference the appropriate APIs.

Lua programming

Lua is a programming language that was developed specifically for expanding applications and to be used in embedded systems. So, many applications support Lua as their plugin programming language of choice. Wireshark is no exception to this.

To learn more about Lua, you can go to `https://www.lua.org/`. You can also go through the various books on Lua programming published by Packt.

You can also go on the Lua website and refer to their documentation section, where you'll find a link `https://www.lua.org/pil/` that covers Programming in Lua:

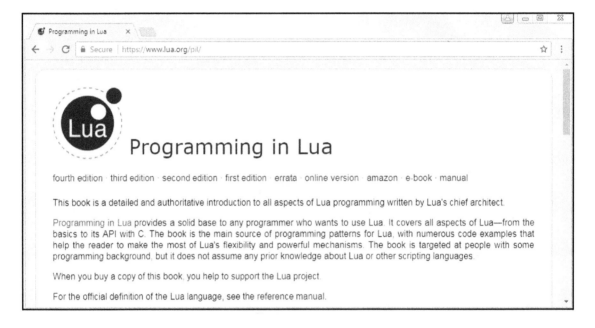

You can see a free book that's available for you to learn programming in Lua. Additionally, you can take a look at the Lua reference manual at `https://www.lua.org/manual/5.1/`:

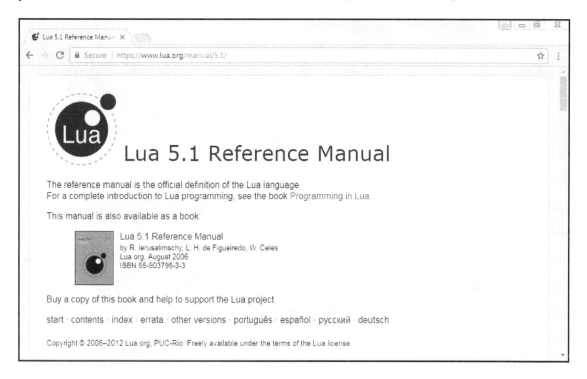

It gives you even more detail as to how Lua works, how the programming language is created, and all the details of how it actually functions, if you need more information than just a standard programming how-to manual.

Once you have downloaded a Lua plugin or you have created your own, you can run it in Wireshark as follows:

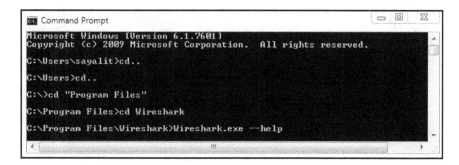

You can then go to where Wireshark is from the command-line and if we run `Wireshark.exe` and you look at the help, you will get the following:

```
Command Prompt                                                    ─  ▭  ⌧

C:\Program Files\Wireshark>Wireshark.exe --help

C:\Program Files\Wireshark>

Wireshark 2.6.0 (v2.6.0-0-gc7239f02)
Interactively dump and analyze network traffic.
See https://www.wireshark.org for more information.

Usage: wireshark [options] ... [ <infile> ]

Capture interface:
  -i <interface>              name or idx of interface (def: first non-loopba
  -f <capture filter>         packet filter in libpcap filter syntax
  -s <snaplen>                packet snapshot length (def: appropriate maximu
  -p                          don't capture in promiscuous mode
  -k                          start capturing immediately (def: do nothing)
  -S                          update packet display when new packets are capt
  -l                          turn on automatic scrolling while -S is in use
  -I                          capture in monitor mode, if available
  -B <buffer size>            size of kernel buffer (def: 2MB)
  -y <link type>              link layer type (def: first appropriate)
  --time-stamp-type <type>    timestamp method for interface
  -D                          print list of interfaces and exit
  -L                          print list of link-layer types of iface and exi
  --list-time-stamp-types     print list of timestamp types for iface and exi
```

You'll see that there is a `-X <key>:<value>`. This is your `eXtension option`, also known as dissectors and plugins.

Now, let's run a Lua script by doing the following:

```
Wireshark.exe -X lua_script:plugins\script.lua
```

So here we are saying: Run an extension in Wireshark. It'll be a Lua script, and it has the filename `script.lua`. Now, it will look wherever you happen to be running the executable, which in this example is under `Program Files\Wireshark`, so you'll want to put your script into the root directory there or in a plugins folder. You could do `pluginsscript.lua` or whatever the name is of your dissector is. Once you do that and push *Enter*, it will run Wireshark and try to load that file.

We will now get into our troubleshooting scenario. We'll determine where to capture our traffic in order to best analyze the data and resolve the issue as quickly as possible.

Determining where to capture

In this section, we'll take a look at determining where best to start a packet capture for the troubleshooting scenario. Now, in this troubleshooting scenario, we have a user that is reporting that they're unable to access the FTP server. They start with their client, and it just says that the connection does not work.

Now, what we need to do is determine where we need to begin packet captures in order to figure out what's going on:

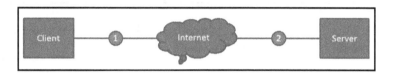

There might be an issue on the client side; there might be an issue on the server side; or there might be an issue somewhere in between on the internet, possibly. Maybe it's a routing issue, or something like that that's out of our control. So besides taking a look at log files on the client or the server, we'll take a look at the packet captures themselves. Additionally, there might be a problem somewhere between the client and the internet, or between the internet and the server. These intermediary devices could be firewalls or routers, or something else that might be blocking or causing issues with the connection. With this scenario, since the client is reporting a problem but is not being very specific, we'll start out with a packet capture on the client side. Then, if we determine that we might need some additional packet captures, we will capture on the server side.

Normally, when you are capturing, you'll start from the most easily accessible location or closest to the issue at hand, and then work your way along the data path. If a client is reporting a problem, then you will most likely be trying to do a packet capture on the client side, and then, if you do not see any sort of obvious issue as to what's going on, then you work your way up. You go to the next intermediary device, such as numbers 1 and 2 on this diagram, and then you go from there; and you might have to get involved with the ISP. Then, you continue your way along that data path from source to destination, performing packet captures to help determine what the issue is.

Capturing scenario traffic

In this section, we'll take a look at capturing some traffic for our troubleshooting scenario, and checking for some obvious issues before we look into the packet capture a bit more in-depth.

In the following screenshot, we have captured the traffic from the client connecting to the server:

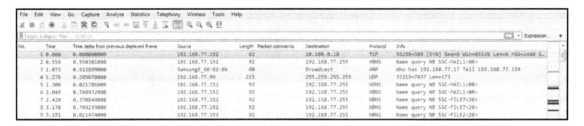

We will now put a filter in here for port `21` because we know that the client is connecting over standard unencrypted FTP. For that we use `tcp.port == 21`:

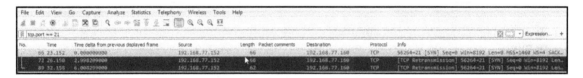

So, we got rid of everything else. We can see that there are three packets, and it looks like we have a `SYN` and two retransmissions. So the client, which is `.152`, is trying to connect to the server running on `.160`, and it's not even beginning the TCP handshake. So the server is not doing something correctly in order to negotiate port `21`. It's not that the server is rejecting the user credentials or there's some sort of other obvious issue. It's definitely not listening on `21` or something like that. Let's start a capture on the server side, and when the client connects, we'll see what it does as well:

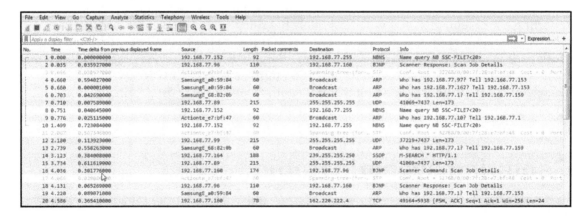

We have the server running, so let's go ahead and try our login attempt. You can see the packets updating and the client's trying to connect, and you can see that it has failed. So we'll stop that and also our capture, and then we'll do what we did earlier.

Let's enter `tcp.port == 21` and see what happens:

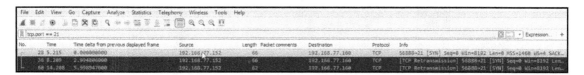

So on the server side, it looks identical. What we're seeing is the source coming into the server as the `152` client, hitting the server's destination IP `160`. It's reaching the server at least, so we know that the IP is working correctly. We'll get a `SYN`, then a retransmission from the client, and then we'll get another retransmission. Obviously, something on the server's not running correctly. Since that's pretty obvious, let's take a look and see whether our FTP server's even running correctly. Let's take a look at our services in here:

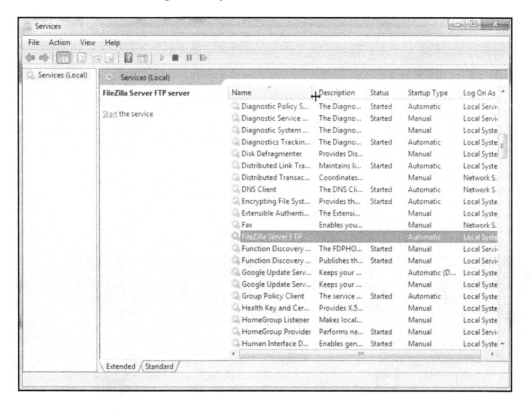

You can see that we have the FileZilla FTP server that's supposed to be running, but it's not started. Let's go ahead and start this service:

Now our server's running and we know that everything looks good; we'll refresh just to make sure; then, let's go ahead and start another capture on the server.

From the client, we will begin a connection, and we can see this:

So we have a SYN, a retransmission again, and another retransmission. The server is running and the service is operational. What we can also do is check our ports that are listening on the server with netstat -an:

```
netstat -an
```

Now, let's take a look at port 21:

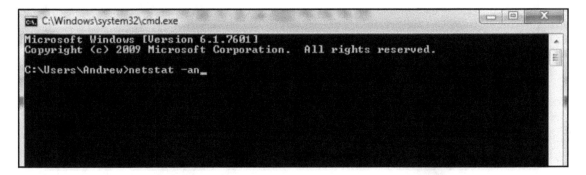

We can see here that we do have port 21 listening on the server, so the service is running correctly. It looks like we'll need to investigate the server a little bit further, and take a look at the packet capture once we eventually get this SYN connectivity issue resolved in the next section.

Now we'll actually perform these packet captures. We'll dive into the diagnostics of our connection issue with FTP here a little bit further.

Diagnosing scenario traffic

In this section, we'll take a look at diving into the captured packets and taking a look at how that FTP server is causing some sort of problem with our connectivity. In our last section, we saw some of the basic settings to make sure that the service was running for the FTP server. We took a capture from the client and saw that there was a TCP SYN with some retransmits, so port 21 was not answering. Then, when that service was up and running, we took a look at the server side again and saw that there was still a TCP SYN with some retransmits. There's still something on the server that's not working correctly, and we know it's on the server side (at least, seemingly) because there's a TCP SYN that is arriving at the server. If it's arriving at the server, it's getting through the firewalls at both ends of the connection; it's getting through the routing on the internet, so we know we're getting partway there, but there's something on the server that's not quite right.

Let's try a connection from the client: just make sure nothing has changed:

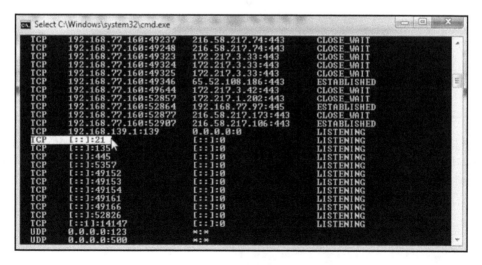

We see our TCP SYN come in, and we have two retransmits from the client. At that point, the client gives up and states that it cannot connect. Since this server is running on Windows, we should take a look at whatever on Windows might be blocking port 21 from responding. So we know the service is running, and we looked at `netstat -an` last time and we saw that port 21 was listening, so that's all good; but something else in the system is blocking it.

If we take a look at the Windows Firewall settings, we see that our firewall is on:

It is most likely that the firewall is blocking it, because the FTP server did not put an exception into the firewall rule. We can create an inbound rule for port 21, or we can just go ahead and disable it for the sake of testing.

For now, we'll just turn this **Off**:

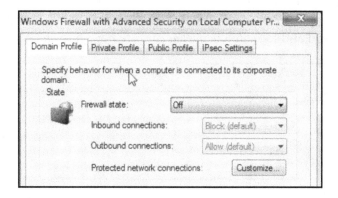

We will do that on all of the profiles.

Now, let's go ahead and do another connection attempt. We will start the connection from the client:

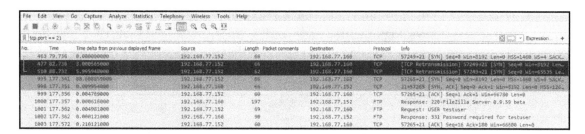

Now, we can see that we have some FTP traffic. So finally, our client is able to connect partially. On the client side, it's currently telling me that it wants a password. We can see that we have [SYN, ACK] and [ACK] for the TCP handshake. Then the server responds, and says it's running the FileZilla server. My client then requests that it logs in with the testuser, and then the server responds, saying that a password is required. My client then acknowledges that request.

So now, let's go ahead and try to log in with the credentials that the client received:

1487 263.841	0.004401000	192.168.77.152	60	192.168.77.160	TCP 57280→21 [ACK] Seq=1 Ack=1 Win=66780 Len=0
1488 263.842	0.000301000	192.168.77.160	197	192.168.77.152	FTP Response: 220-FileZilla Server 0.9.59 beta
1489 263.847	0.004803000	192.168.77.152	69	192.168.77.160	FTP Request: USER testuser
1490 263.847	0.000130000	192.168.77.160	90	192.168.77.152	FTP Response: 331 Password required for testuser
1491 264.054	0.207360000	192.168.77.152	60	192.168.77.160	TCP 57280→21 [ACK] Seq=16 Ack=180 Win=66600 Len=0

You can see at the bottom that we've actually got `testuser` and password. Let's log in with the credentials that the client was provided with. You will get a access denied error, so the client still can't connect correctly. After fixing two different things, the client still cannot connect. Let's go ahead and look in our packet capture here, and see if we can find out why:

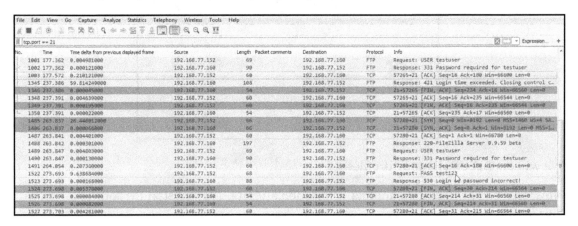

So we have our `SYN`; `SYN`, `ACK`; and `ACK`—the server responded. We tried to log in with `testuser`. The server says: log in with a password. We acknowledged that. Then the client provides a password `test123`. Then the server states that the login is incorrect. What we'll need to do is reset the password in the server:

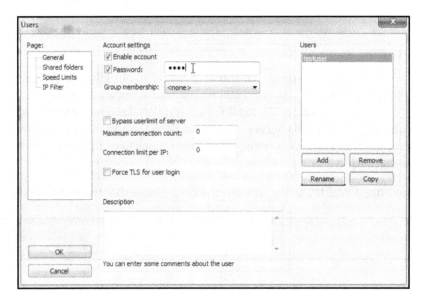

You can see on to the server running FileZilla. Go ahead into the **Users** section, and we will change the password on this to `test123`, which is what the client is expecting it to be. Let's go ahead and try that connection again:

1900	349.592	0.004447000	192.168.77.152	60	192.168.77.160	TCP	57297→21 [ACK] Seq=1 Ack=1 Win=66780 Len=0
1901	349.592	0.000323000	192.168.77.160	197	192.168.77.152	FTP	Response: 220-FileZilla Server 0.9.59 beta
1902	349.597	0.005048000	192.168.77.152	69	192.168.77.160	FTP	Request: USER testuser
1903	349.597	0.000157000	192.168.77.160	90	192.168.77.152	FTP	Response: 331 Password required for testuser
1904	349.602	0.004690000	192.168.77.152	68	192.168.77.160	FTP	Request: PASS test123
1905	349.602	0.000181000	192.168.77.160	69	192.168.77.152	FTP	Response: 230 Logged on
1906	349.607	0.004726000	192.168.77.152	60	192.168.77.160	FTP	Request: SYST
1907	349.607	0.000122000	192.168.77.160	86	192.168.77.152	FTP	Response: 215 UNIX emulated by FileZilla
1908	349.612	0.004662000	192.168.77.152	60	192.168.77.160	FTP	Request: FEAT
1909	349.612	0.000085000	192.168.77.160	176	192.168.77.152	FTP	Response: 211-Features:
1910	349.617	0.005108000	192.168.77.152	81	192.168.77.160	FTP	Request: CLNT WinSCP-release-5.9.3
1911	349.617	0.000094000	192.168.77.160	70	192.168.77.152	FTP	Response: 200 Don't care
1912	349.622	0.005037000	192.168.77.152	68	192.168.77.160	FTP	Request: OPTS UTF8 ON
1913	349.622	0.000122000	192.168.77.160	118	192.168.77.152	FTP	Response: 202 UTF8 mode is always enabled. No need t..
1914	349.704	0.081625000	192.168.77.152	60	192.168.77.160	FTP	Request: PWD

If we take a look here, we can see that we need `testuser` and password. The `test123` password is sent. The server states that it has now logged on—so that was a successful login with the correct credentials. Then, go ahead and continue with the additional commands that we referenced in the prior sections. What we'll do then is disconnect:

2182	412.503	2.462598000	192.168.77.152	60	192.168.77.160	TCP	57297→21 [FIN, ACK] Seq=121 Ack=669 Win=66112 Len=0
2183	412.503	0.000034000	192.168.77.160	54	192.168.77.152	TCP	21→57297 [ACK] Seq=669 Ack=122 Win=68560 Len=0
2184	412.503	0.000079000	192.168.77.160	54	192.168.77.152	TCP	21→57297 [FIN, ACK] Seq=669 Ack=122 Win=68560 Len=0
2185	412.508	0.004683000	192.168.77.152	60	192.168.77.160	TCP	57297→21 [ACK] Seq=122 Ack=670 Win=66112 Len=0

We can see that from the client we initiated a finalization, a `FIN`; `ACK`; `FIN` from the server; and acknowledge. We've got the four-way handshake to terminate the connection.

Summary

In this chapter, we were able to successfully diagnose multiple issues with this server. Initially, the service was not running. Then, the service was being blocked by the Windows Firewall. Then, there were incorrect credentials provided to the client. Furthermore, using Wireshark we were able to diagnose and provide some information as to where we needed to look, whether it was on the client side or the server side, or in between, and it was a great help, allowing us to resolve this issue quickly. In this chapter, we talked about Wireshark plugins, using Lua and creating dissectors. We then evaluated a troubleshooting scenario, determining where to create our captures, capturing the traffic, and doing some basic diagnostics, and then diving into it a little bit more and actually reading the FTP packets, once they started to send and receive to the client. We have now reached the end of the book. I hope you have enjoyed reading it.

Other Books You May Enjoy

If you enjoyed this book, you may be interested in these other books by Packt:

Network Analysis using Wireshark 2 Cookbook - Second Edition
Nagendra Kumar Nainar, Yogesh Ramdoss, Yoram Orzach

ISBN: 978-1-78646-167-4

- Configure Wireshark 2 for effective network analysis and troubleshooting
- Set up various display and capture filters
- Understand networking layers, including IPv4 and IPv6 analysis
- Explore performance issues in TCP/IP
- Get to know about Wi-Fi testing and how to resolve problems related to wireless LANs
- Get information about network phenomena, events, and errors
- Locate faults in detecting security failures and breaches in networks

Cybersecurity – Attack and Defense Strategies

Yuri Diogenes, Erdal Ozkaya

ISBN: 978-1-78847-529-7

- Learn the importance of having a solid foundation for your security posture
- Understand the attack strategy using cyber security kill chain
- Learn how to enhance your defense strategy by improving your security policies, hardening your network, implementing active sensors, and leveraging threat intelligence
- Learn how to perform an incident investigation
- Get an in-depth understanding of the recovery process
- Understand continuous security monitoring and how to implement a vulnerability management strategy
- Learn how to perform log analysis to identify suspicious activities

Leave a review - let other readers know what you think

Please share your thoughts on this book with others by leaving a review on the site that you bought it from. If you purchased the book from Amazon, please leave us an honest review on this book's Amazon page. This is vital so that other potential readers can see and use your unbiased opinion to make purchasing decisions, we can understand what our customers think about our products, and our authors can see your feedback on the title that they have worked with Packt to create. It will only take a few minutes of your time, but is valuable to other potential customers, our authors, and Packt. Thank you!

Index